Water-Efficient Landscaping
in the
Intermountain West

Water-Efficient Landscaping in the Intermountain West

A Professional and Do-It-Yourself Guide

Heidi A. Kratsch, Editor

Utah State University Press
Logan, Utah
2011

Utah State University Press
Logan, Utah 84322-7800
www.usu.edu/usupress

Manufactured in the United States of America

ISBN: 978-0-87421-789-6 (paper)
ISBN: 978-0-87421-790-2 (e-book)

Library of Congress Cataloging-in-Publication Data
Water-efficient landscaping in the Intermountain West : a professional and do-it-yourself guide / Heidi A. Kratsch, editor.
p. cm.
Includes bibliographical references.
ISBN 978-0-87421-789-6 (pbk.) — ISBN 978-0-87421-790-2 (e-book)
1. Landscape architecture in water conservation—Great Basin—Handbooks, manuals, etc. 2. Water efficiency—Great Basin—Handbooks, manuals, etc. 3. Irrigation efficiency—Great Basin—Handbooks, manuals, etc. 4. Drought-tolerant plants—Great Basin—Handbooks, manuals, etc. 5. Xeriscaping—Great Basin—Handbooks, manuals, etc. I. Kratsch, Heidi A.
SB475.83.W364 2011
635.9'5250979—dc22

2010052096

Table of Contents

Foreword

Utah is one of the five fastest growing states in the nation, and it shares a border with the four other fastest growing states. From 1990 to 2000, Utah's population increased by more than 510,000 people to over 2.2 million. In simple terms, Utah's semiarid terrain is sprouting another city approximately the size of Salt Lake City (1990 population of 160,000) about every three years.

According to the Governor's Office of Planning and Budget, Utah's rapid growth will continue, with the population more than doubling to nearly five million by 2050. As Utah's population blossoms, so will the demand for Utah's limited water resources. If Utah's municipal and industrial (M&I) water demands increase at the same rate as its population, the state is headed for trouble. However, since Utahns currently use more water than they need, particularly in watering their landscapes, the opportunity exists to avoid many of these problems by reducing use to a more efficient level.

Water conservation will play a significant role in helping Utahns use water more efficiently. This increased efficiency will reduce water demands and help water suppliers meet future needs. In addition to this important benefit, water conservation can also:

- Delay expensive capital investments to upgrade or expand existing water facilities.
- Reduce sewage flows, delaying the need for more wastewater treatment facilities.
- Conserve energy, as less water needs to be treated, pumped, and distributed to the consumer.
- Lessen the leaching of chemicals and sediments into streams and aquifers.
- Reduce stream diversions, thereby enhancing water quality as well as environmental and recreational functions.
- Improve water levels in reservoirs.

Few Utahns dispute the value of water conservation; however, there is debate about how it will happen and who is responsible for making sure it does. In reality, every Utahn is responsible for making it happen.

The state of Utah has developed a specific goal to conserve water use directly linked to M&I needs. This goal is to reduce the 1995 per capita water demand from public community systems by at least 25% before 2050. Specifically, demand will need to decline from 321 gallons per capita per day (gpcd) to a sustained 240 gpcd or less. The accomplishment of this goal is equivalent to a total decrease in demand of about 400,000 acre-feet per year by 2050, and represents the most significant component in meeting Utah's future water needs.

The goal is based on modeling and research that indicates indoor and outdoor water use can reasonably be reduced by 25% or more. Indoor reductions will be realized through public education and the installation of more efficient fixtures and appliances. Outdoor reductions will also be realized through public education, along with more efficient application of water on landscapes, and reduction in turfgrass areas. Water-wise landscaping will be an integral part of meeting our goals.

Eric Klotz
Utah Division of Water Resources

Preface

This manual provides information on the technical aspects of designing, building, and maintaining water-efficient landscapes in the intermountain region of the western United States. It was written specifically for professionals: landscape designers, architects, contractors, maintenance professionals, and irrigation specialists who desire to create landscapes that are both water-sustainable and beautiful. For this reason, we have assumed the reader has some basic knowledge about landscape and irrigation design, plant materials, and landscape maintenance practices. However, this manual may also be of interest to plant growers and retailers, botanical garden staff, educators, city and state government personnel, water agencies, and do-it-yourself homeowners who wish to learn about designing and maintaining successful water-efficient landscapes. For those who require more basic information about a specific topic, we have included a list of suggested references at the end of the book.

This is meant to be a working manual. In other words, we don't expect the reader to start at the beginning of the manual and read straight through to the end, although that would certainly be helpful. We realize that many readers will have a greater interest in their own area of specialty, and may want to focus on those chapters that are of most immediate use. We also realize that you, our reader, may be like us— pressed for time and forced to get your information on a need-to-know basis. We have designed this manual with you in mind. It is possible to go directly to any chapter in the manual and be able to implement the principles and practices discussed. Chapters are written in a simple and easy-to-read style, with clear headings and bullet points to help you access specific areas quickly. Within each chapter, we reference other chapters that provide more detailed information on a related topic. In this way, you may end up reading the whole manual eventually, but you will be able to refer to specific chapters according to your needs and time constraints. For your convenience, the manual has a sewn, lay-flat binding. This will allow you to take it with you on-site if you desire, and to make copies of some of the pages that you refer to often. Other pages may be especially useful for your clients, and you are free to make copies for their use; for example, the Irrigation System Maintenance Checklists in "Principle 6: Irrigation Planning" would be a helpful tool to remind clients about irrigation system maintenance.

And speaking of chapters—we have arranged this manual in a nontraditional format, calling the chapters "principles." This was intentional. Each chapter represents and explains in detail how to implement one of the seven principles of water-efficient landscaping. We hope, even if you never read the entire manual, that you will at least know and understand the "seven principles." They are the keys to understanding and using the concept of water-efficient landscaping. We sincerely want to help you succeed—understanding and designing your landscapes with all seven principles in mind will enable you to know at a deep level how water-efficient landscapes function, and will also help your clients to know this too so they may enjoy their new landscape for many years to come.

As you can see by the long list of authors, this manual is the culmination of many hours of work by a lot of people. I thank my coauthors for their patience and persistence in coming together to work out a common vision, and then writing their chapters with that vision in mind.

Heidi A. Kratsch

Principle 1

Planning and Design

Anne Spranger and William A. Varga

The design of a landscape is critical to achieving water sustainability. Landscapes must be designed to match the regional climate and soil conditions. These factors determine the amount of supplemental water that will be needed at a site. Ideally, the landscape should be designed so that the right plants are planned for the right spots, and plants with similar needs are grouped together so the need for supplemental water is minimized.

The first step in achieving this goal is to have a well-thought-out plan.

Program Development

Program development is an important first step in designing a successful water-efficient landscape. This step involves meeting with the client and discussing their landscape needs. This may include both the aesthetics and functional requirements for the landscape. This is a written exercise. Talk to the client and make a list of landscape goals and objectives. Determine activities, use areas, circulation elements, and related facilities that are desired in the final landscape. The available budget and desired water savings should be included in this discussion. The following are topics for consideration during a client interview:

- ACTIVITIES AND USE AREAS: What activities are planned for this landscape? Activity areas for a residential homeowner might include space for outdoor dining and barbecuing, children's play, vegetables and herbs, specialty gardens, a fruit/ nut tree grove, pets, storage, and composting.

Limit turfgrass to areas where it provides a functional benefit.

- CIRCULATION: How will the flow of traffic through this landscape be controlled? Circulation elements define the flow of traffic through a landscape. They should be both functional and aesthetically pleasing. Evaluate existing conditions and future needs for automobile access and parking, front entry walk design, and all secondary or tertiary paths.
- HARDSCAPE: Evaluate existing conditions and future needs for hardscape elements such as patios, decks, courtyards, and shade structures. How do hardscapes in this landscape connect with planted areas?
- IRRIGATION: Does the client want an automatic irrigation system? If so, how much of the budget can be allotted to this? Is the client interested in water harvesting: roof collection systems, rain barrels, or underground water-collecting cisterns? These systems allow a homeowner to collect precipitation water and either divert it to a planted area or store it for later use on the landscape. They conserve water by preventing evaporation or runoff from the site. (Check with your local county planning office to ask about zoning ordinances for water rights issues.)
- PLANT SELECTION: Does the client have favorite plants, favorite colors, plant allergies? Does the client desire a lawn? Determine planned use for a lawn and how big it needs to be. Limit turfgrass to areas where it provides a functional benefit.
- SPECIAL CONSIDERATIONS: Does the client have unique needs that need to be considered? Talk to the client about various topics such as desired maintenance levels, landscape lighting, ADA compliance, and ages of family members.

There are many other subjects that can be discussed with the client (see the Client Questionnaire Checklist at the end of this chapter). When the designer understands the client's needs, the resulting water-efficient landscape will be more successful.

Landscape Styles

A water-efficient landscape can be any style. It can be formal or informal, historic or modern, native or non-native; it can even have a theme if desired, such as a children's garden, a fragrance garden, or a pollinator garden. In general, landscape style should be influenced by the region and climate of the site, the architecture of the buildings, and of course, the client's preference. A combination of landscape styles may also be used if one style does not suit the site or client.

Photo and landscape design: Anne Spranger

Figure 1-1. A raised-bed kitchen garden is a themed garden with both aesthetic and functional appeal. A variety of herbs, small fruits, and vegetables can be grown. They will add diversity to the landscape and provide a source of fresh produce throughout the growing season. Utah House, Kaysville, UT

Formal and Informal Landscapes

A *formal* water-efficient landscape may have the appearance of a formal conventional landscape, which suggests a control of nature through its orderly, symmetrically balanced grouping of elements. Plants are often used to reflect the geometry of building architecture or the lines of streets and walkways. Planting areas and circulation elements are often straight and rigid. Plants are generally chosen for their formal, stiff, or manicured appearance. For example, formal landscapes may have edged lawns and neatly hedged shrubs. Keep in mind that formal landscapes often require a higher maintenance level than do informal landscapes. Clients should be advised about this; if they are unwilling or

Figure 1-2. A semi-formal landscape includes elements of both formal and informal designs. In this low water-use semi-formal landscape, the lawn is neatly mowed and edged in a rectilinear fashion. The perennial beds include a variety of low water-use plant species, which are maintained to enhance their natural growth form. Logan, UT.

unable to perform these more demanding maintenance tasks, they should consider whether hiring labor for landscape maintenance is within their budget.

On the other hand, an *informal* water-efficient landscape is less geometric and uses asymmetrical balance. Planting areas and circulation elements tend to be free-flowing, incorporating curvilinear or irregular shapes and lines. Plants in an informal landscape may be maintained to accentuate their natural form.

Naturalistic Water-Efficient Landscapes

A *naturalistic* water-efficient landscape is designed using a very different approach from that of a formal or informal design. While form, color, and texture are still important for plant selection, the plants in a naturalistic design are generally native plants, and are grouped according to their natural associations or

Photo and landscape design: Anne Spranger

Figure 1-3. An informal landscape is designed with asymmetrical balance, and incorporates curvilinear shapes within circulation and hardscape elements. The plants in this landscape are low to moderate water-use. Utah Botanical Center Rasmussen Teaching Garden, Kaysville, UT.

plant communities. Creating a sense of place and a sustainable landscape are of primary importance. A water-efficient naturalistic design for the Intermountain West could mimic natural plant communities in the region such as foothills, mountains, prairies, valleys, or deserts.

Maintenance for a naturalistic landscape is very different than it is for a formal or informal water-efficient landscape. Perfectly manicured lawns and sheared hedges are not found in the natural landscape, and are therefore not encouraged in a naturalistic design style. Spread of perennial plants in a naturalistic landscape may be desirable and promoted, and plants are often chosen for their capacity to provide habitat for wildlife and pollinators. Recycling nutrients by way of composting discarded plant parts and meeting water requirements by limiting water use to natural precipitation are sustainable methods for maintaining a naturalistic water-efficient landscape.

Photo: Anne Spranger. Landscape design: Darrel Morrison, William Varga, and Anne Spranger

Figure 1-4. A naturalistic landscape usually consists of plants native to a region. Plants are selected to reflect regional native plant communities and often provide food and shelter for wildlife and native pollinators such as birds, bees, and butterflies. Utah Botanical Center parking lot, Kaysville, UT.

Site Inventory and Analysis

The second step in designing a successful water-efficient landscape is to become familiar with the site in order to determine its opportunities as well as its constraints. The following types of questions should be answered:

- What types of soil, drainage, exposure, and climatic conditions exist on the property?
- Are there areas that require screening, either from the sun or wind, or to conceal poor views?
- Where are the hot, dry areas on the site?
- What kinds of plants will grow well on the site?
- Do steep slopes present erosion or runoff problems?
- What areas of the site are well suited to accommodate the client's needs and activities?

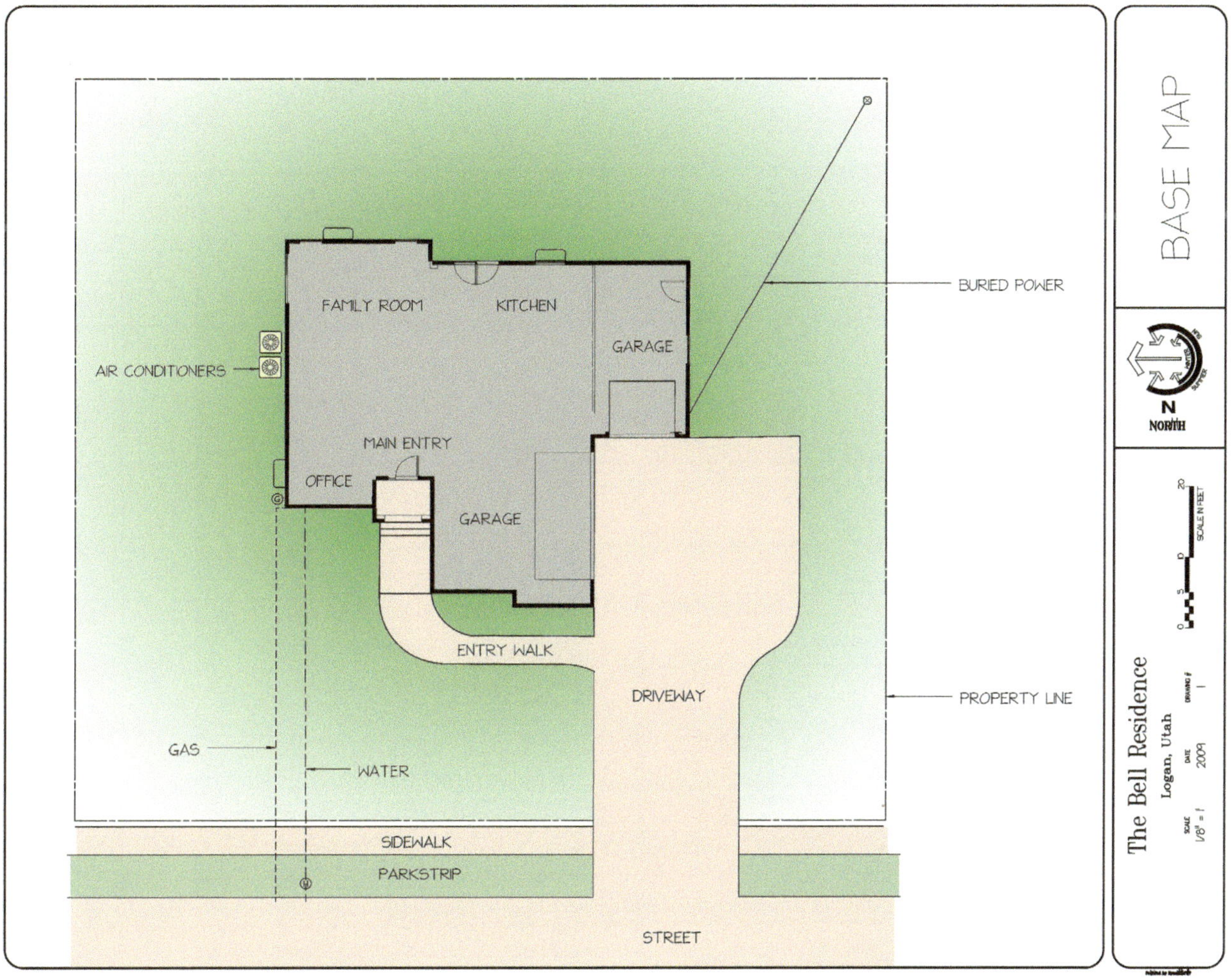

Figure 1-5. Base map.

All Landscape designs: Anne Spranger

- What is the best way to conserve water on this site?
- Where and how is water being wasted on the site? Current water waste and its causes (see "Principle 6: Irrigation Planning")

Start with a scale drawing of the property to be landscaped. Take note of all existing site conditions. Important landscape elements to inventory and analyze include the following:

- SOILS: Determine soil type and its water-holding capacity.
- TOPOGRAPHY: Determine percent slope, drainage, and erosion potential.
- VIEWS: Take note of desirable and undesirable views from both inside the building and from the neighbors' point of view.
- EXISTING VEGETATION: Determine what existing plants will be kept, moved, or removed. Determine weed abatement strategies.
- CLIMATIC INFLUENCES: Determine high and low temperatures, annual precipitation, prevailing wind directions, and frost pockets.

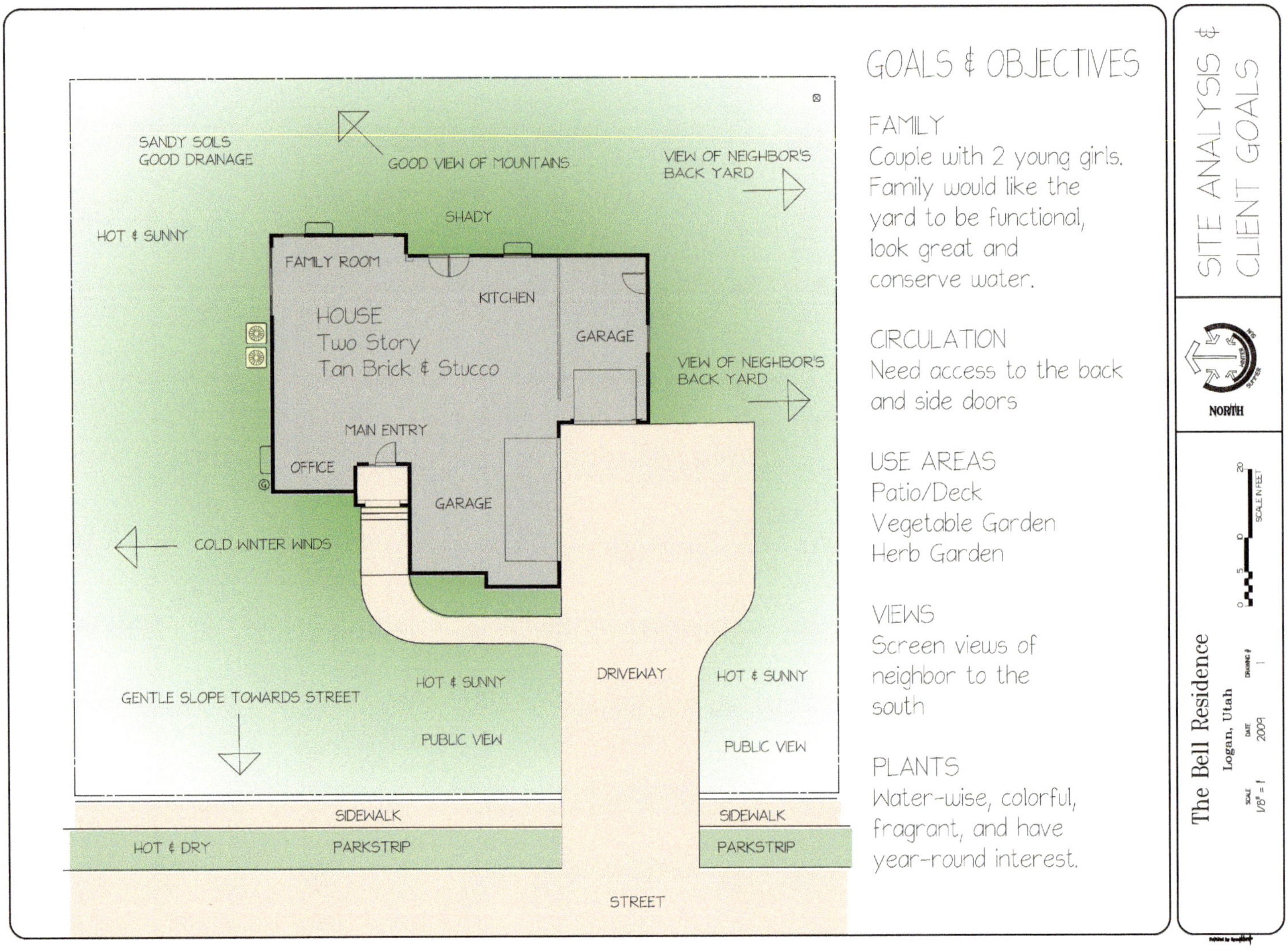

Figure 1-6. Site and family needs analysis.

- EXPOSURE: Determine where sunny and shady areas are in the landscape.
- HARDSCAPE: Take note of all existing hardscapes. Note materials, condition, and design. Determine what to keep and what to remove.
- UTILITIES: Have all utilities marked before digging begins. Utilities to map include power, gas, water, sewer, telephone, and cable. Note whether utilities are buried or overhead, and note buried depth or overhead height.
- ADJACENT PROPERTY: Take note of views, erosion potential, excess water runoff, or any other special conditions that may affect the client's landscape.

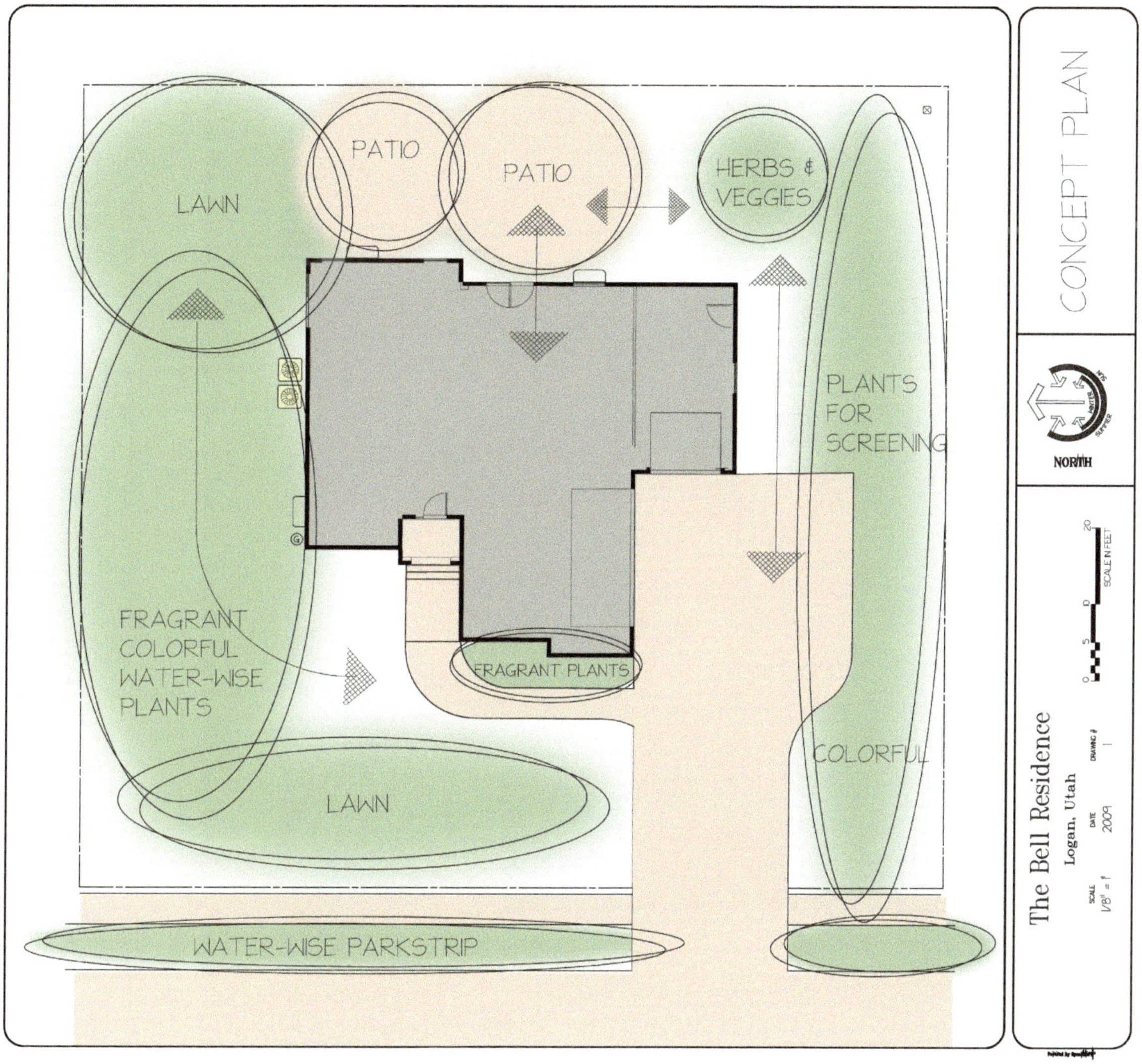

Figure 1-7. Concept plan.

Take special care to consider all factors that contribute to dryness, including soil type, soil drainage, sun exposure, and climatic influences. Designing a successful water-efficient landscape begins by knowing what you have to work with.

Concept Plan

The third step in designing a successful water-efficient landscape is to start designing. At this point, the designer has collected information about the site and the client's goals and objectives. Next, the designer develops a concept plan. The purpose of a concept plan is to study appropriate locations and sizes for the various activities, use areas, and circulation elements that were identified in

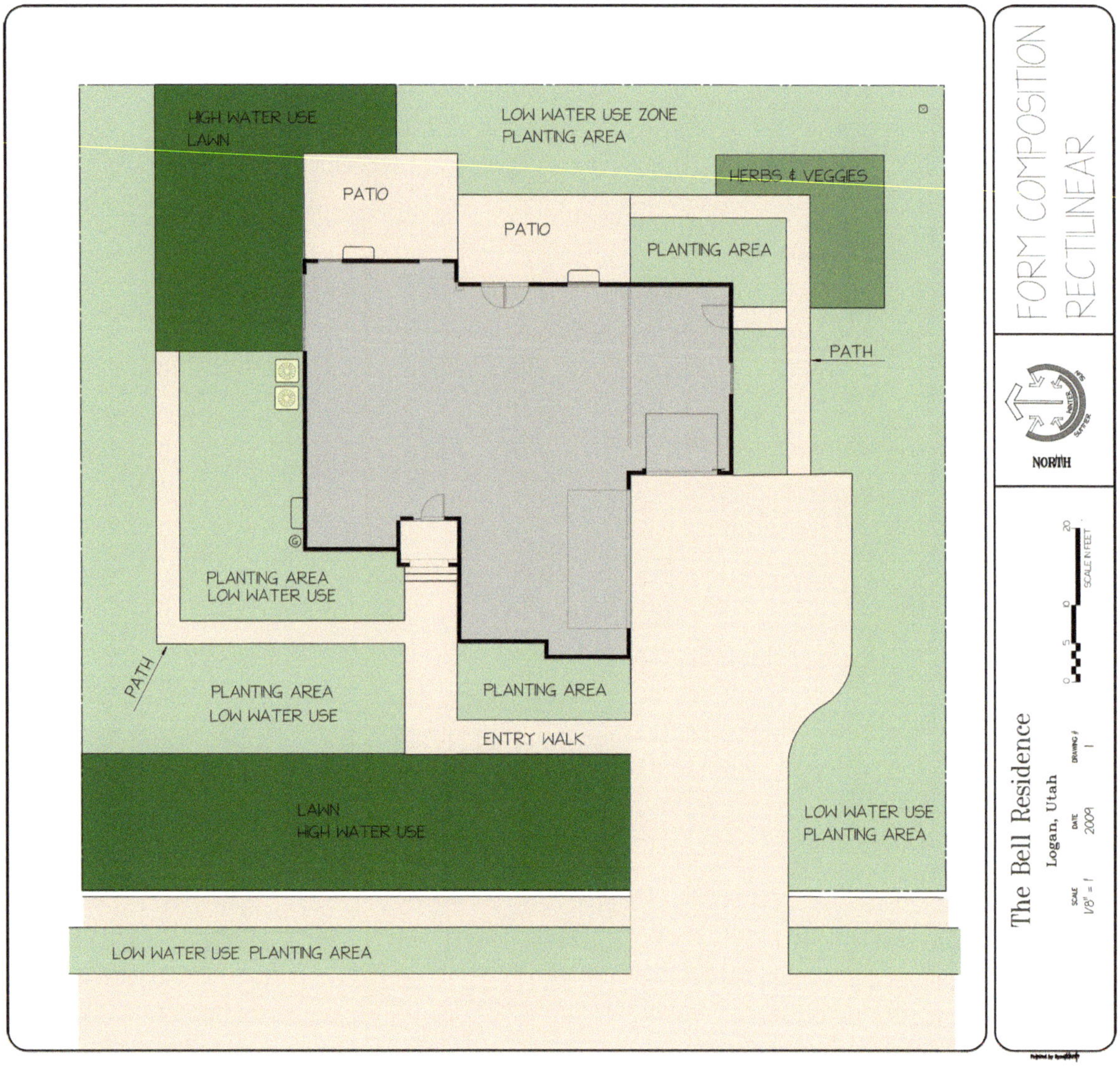

Figure 1-8. Form composition, rectilinear.

the program development and site analysis. A concept plan is drawn freehand; bubbles of various shapes and sizes are used to broadly depict the functional areas within a landscape and how they relate to one another. These drawings are meant to be rough estimates of the use of space at a site. Several concept plans should be drawn in order to find the best solution for a particular site and client. Once a concept plan is finalized, the use areas and circulation elements are converted into specific ground pattern shapes. Ground pattern is also commonly referred to as form composition.

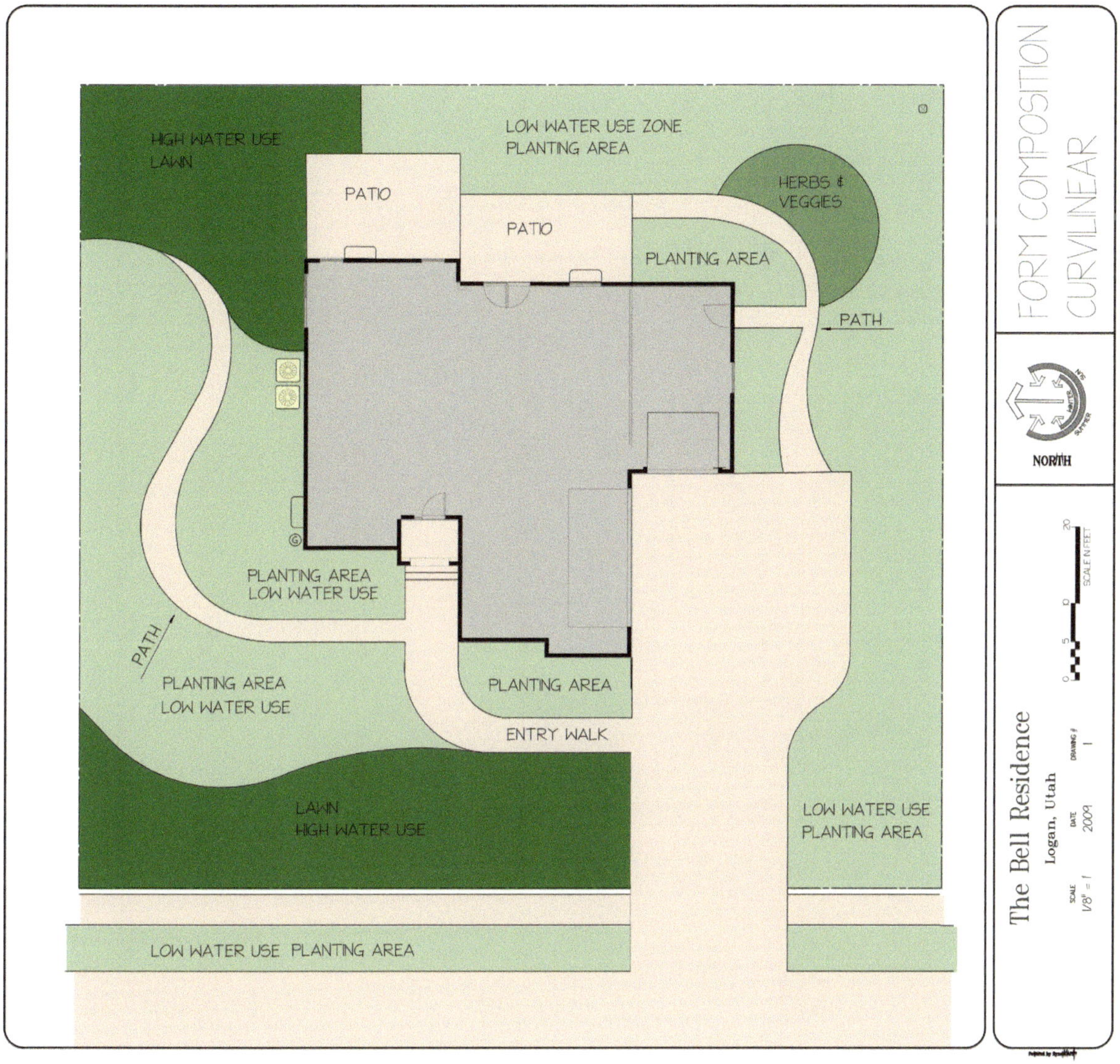

Figure 1-9. Form Composition, curvilinear

Form Composition (Ground Pattern)

Form composition involves creating a specific edge, shape, form, and location for all spaces generated in the concept plan phase. For example, an area that represents the patio on the concept plan is now ready to be given a specific shape. It could be given a formal rectangular shape or perhaps a more informal curvilinear shape. The lawn, planting beds, garden paths, walkways, patios, and decks need to be given a definitive shape.

Form composition also helps to create consistency and continuity throughout a landscape by the repetition of particular forms. These forms or shapes may be angular, rectilinear, curvilinear, arc/tangent, or radial concentric. The choice of

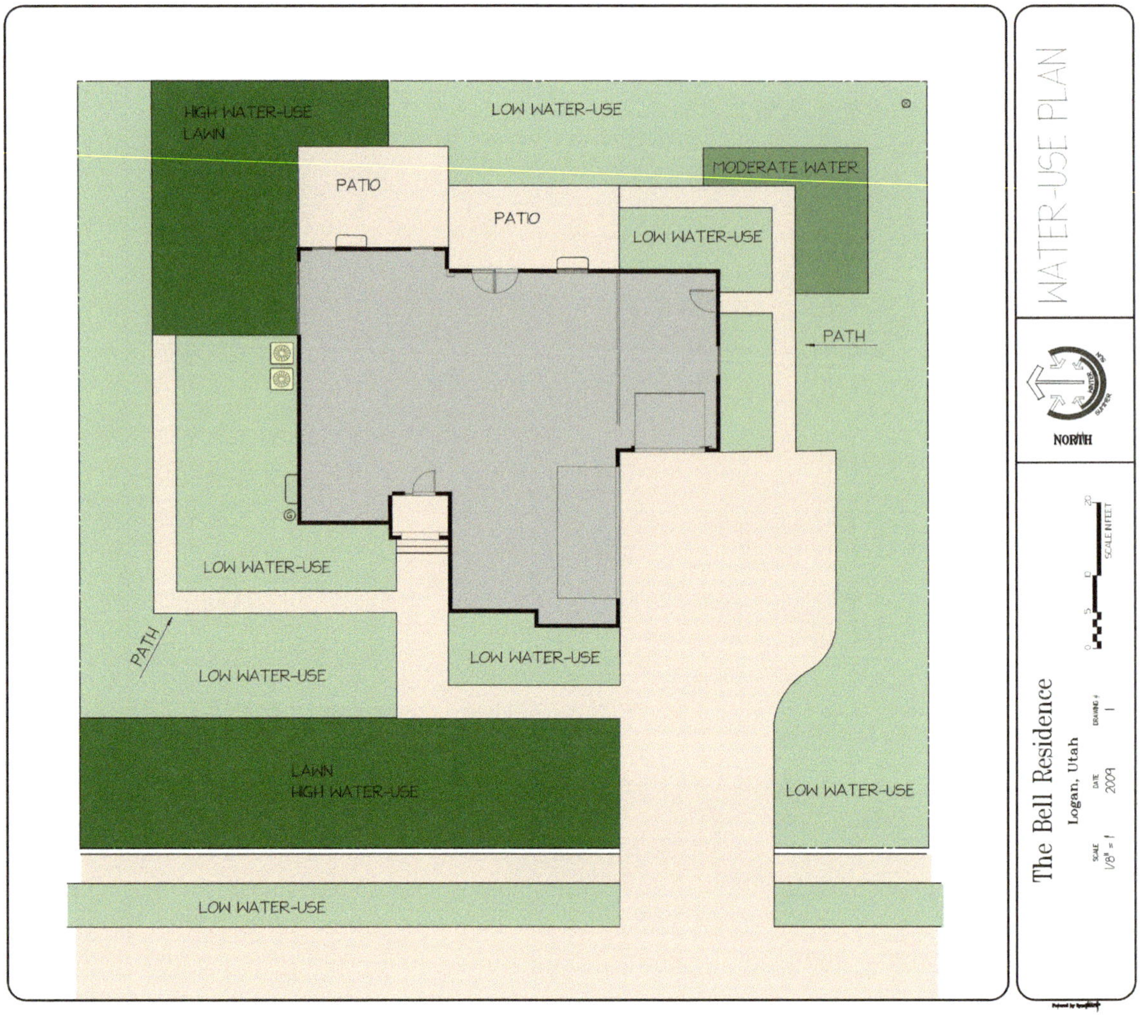

Figure 1-10. Water needs.

specific ground patterns may be based on the desired garden style or character, the character of the site, or the client's personal preference. Defining the form composition is a crucial step in the design process because it directly affects the visual appeal of a space. Visual appeal is enhanced by considering the geometric relationships between linear and circular elements within a form composition. The landscape designer should create at least two or three different ground pattern options for the client to review. Form composition will also influence maintenance and irrigation issues. A complex lawn shape can be challenging to mow and very difficult to irrigate efficiently.

Water-Use (Hydrozone) Zone Plan

It is essential to create a water-use plan for a water-efficient landscape. Water-efficient landscaping takes advantage of the varying conditions and microclimates that exist in every landscape. A microclimate is a small area in a landscape with a climate that differs from that of the surrounding area due to differences in sun or wind exposure, elevation, or soil characteristics. Use healthy, well-adapted plants, and *hydrozone* the landscape by grouping plants together that have similar water needs. It is important to place the right plant in the right location. Avoid planting a high-water-use plant next to a low-water-use plant. Many plants may be used in more than one water-use zone. Site-specific soils will affect water use. Heavy clay soils will require less frequent watering, while sandy soils may require more frequent watering.

Water-Use Zones (Hydrozones)

Very Low Water

This is the lowest-water-use zone in a water-efficient landscape and will provide the greatest water savings when compared to conventional landscaping. Plants used in this zone generally need no additional water once established. Supplemental water may be necessary only during extreme drought conditions, and this could be provided manually. This zone is often placed farthest from the residence because it does not need connections to irrigation pipe to support it.

Low Water

Plants used in the low-water-use zone need more water than the natural precipitation our region provides. Approximately one-half inch of water supplied once or twice per month should be sufficient. This amount may vary from site to site. Careful selection of a low-volume irrigation system will conserve large amounts of water (see “Principle 6: Irrigation Planning”). However, many people prefer to conserve additional water and water this zone by hand, as needed. Low-water-use zones should be as large as possible in order to maximize water conservation benefits.

Moderate Water

Moderate-water-use plants generally need about three-quarters of an inch of water supplied once per week. This is more than the low-water-use plants require but still 50% less than that required by most conventional landscapes.

Figure 1-11. Planting plan.

Remember that site-specific soils may affect water requirements. Plants in heavy clay soils may need less frequent watering, and those in sandy soils may need more frequent watering.

High Water

The high-water-use zone should be kept small in size and functional in use. It could be considered a mini-oasis and is best used when incorporated carefully into your landscape design. Kentucky bluegrass lawn is generally the primary plant used in this zone. Many annual flowers may also be considered high-water-use. Approximately one-half inch of water supplied three times per week is necessary. The high-water-use zone should be located close to the residence to minimize the length of pipe needed for irrigation.

Utah House, Kaysville, UT. Photo and landscape design: Anne Spranger

Figure 1-12. Water-efficient landscapes can be both beautiful and functional. The informal low-water-use landscape surrounding the Utah House in Kaysville, UT is lush and colorful throughout the growing season. The plants were selected for their capacity to moderate seasonal temperatures by creating shade and buffering the wind. They also keep precipitation onsite by preventing runoff of water and erosion of soil.

Planting Design Plan

Landscape design combines elements of art and science to create a functional and aesthetically pleasing outdoor living space. The water-efficient landscape designer must understand the principles of artistic design as they apply to both plant materials and hardscape solutions. Many beautiful trees, shrubs, perennials, and ground covers are available for use in a water-efficient landscape. When selecting plants for the water-efficient landscape, the following characteristics should be considered:

- Cold hardiness
- Water use
- Drought tolerance
- Resistance to pests and disease
- Mature size
- Color

- Form
- Texture
- Leaves
- Flowers
- Fruit
- Year-round interest
- Maintenance needs
- Availability
- Cost

Plant Function

Plants function in many different ways in the designed water-efficient landscape. They may serve as a structural architectural component, an engineering component, or as climate control. Of course, the beauty and aesthetics of plants should be considered as well.

Plants and Structure

Plants can be used to give structure to a water-efficient landscape by defining spaces and creating outdoor "rooms." Trees and shrubs may be used to create walls, screens, or barriers. Ground covers and turfgrass can be used to create the floor of an outdoor room, while trees can create the overhead plane or ceiling.

Plants and Environmental Functions

If placed strategically, plants can be used as environmental controls. They can modify the climate by buffering wind, modifying temperatures, creating shade, and filtering dust and noise. Plants also serve as a means to minimize glare, reduce air pollution, and curb potential erosion.

Plants and Aesthetics

It is an unfortunate misconception that all water-efficient plants are thorny and colorless. In actuality, water-efficient plants come in all colors, sizes, textures, and shapes. It is essential for the water-efficient landscape designer to have thorough knowledge of the large array of beautiful plants available for the water-efficient landscape (see "Principle 3: Plant Selection").

Physical Properties of Plants

Form

It is essential for the landscape designer to understand plant form for both functional needs (shade, screening, wind protection, enclosure) and for aesthetic reasons. General plant forms for trees include round, oval, columnar, vase-shaped, weeping, and pyramidal. Shrubs may be round, oval, upright, pyramidal, prostrate, or creeping. Flowers also come in many different forms. Again, it is crucial for the designer to understand and properly use the various plant forms in order to create a successful landscape design.

Color

A common misconception regarding water-efficient landscapes is that they are not colorful. In actuality, water-efficient gardens can be extremely colorful. A well-designed water-efficient garden will have color throughout all seasons, from early spring to late fall, and even throughout the winter. Foliage color is present during most of the year, so it must be considered in the design process. Flowers, fruit, bark, and seeds also produce colors worthy of consideration.

Color is often used to direct attention in the water-efficient landscape; therefore, it must be used carefully. The landscape designer should consider the color of the home or building as well as any outdoor hardscape colors used on the project site. Colors can be used to visually change perspective in the landscape. Warm colors, such as red, orange, and yellow, are bright, inviting, and lively. Warm colors tend to advance an area toward an observer, thus making a large space appear smaller or more intimate. Cool colors, on the other hand, such as blue, green, violet, and pink, tend to recede into the landscape. These colors are regarded as calm and restful. When used properly, cool colors can make a small space appear larger.

Texture

Texture is very important in the water-efficient landscape. It describes the surface quality of an object that can be seen or felt. Textures of landscape plants can be coarse, medium, fine, smooth, rough, glossy, dull, fuzzy, soft, or even prickly. Surfaces in the landscape that have texture include buildings, walks, patios, and plants. Leaves, bark, twigs, branches, and buds are the physical features that determine the texture of a plant. In general, fine-textured plants appear lacy, feathery, airy, or light, and have small foliage on delicate stems. Coarse-textured plants have large leaves and stiff, sturdy stems. For example, a fern leaf has a fine, soft, feathery texture, while a yucca plant has leaves that are large, bold, coarse, and sharp. Hardscape building materials have varying textures as well. Brick often has a coarse texture, while vinyl siding is smooth. The landscape designer must understand how to find the right balance of textures for each design plan. Therefore, it is essential for the landscape designer to have a thorough knowledge of the proposed plant material and all hardscape building materials.

Utah House, Kaysville, UT. Photo and landscape design: Anne Spranger

Figure 1-13. The parking strip in front of the Utah House was designed to provide all-season appeal. These pictures show how the landscape looks in spring (above) and in late summer (facing page).

Hardscape Design

An obvious way to reduce water use in the home landscape is to incorporate hardscape design. A well-planned water-efficient landscape usually has several hardscape features, including patios, decks, walkways, paths, and driveways. Hardscape features can add function, beauty, increased outdoor living space; and best of all, they don't require irrigation. Do not make the mistake, however, of substituting hardscape for plants. Appropriately selected plants conserve more water than hardscapes do because plants cool the air around the home by evaporating water through their leaves and minimizing heat- and light-reflective surfaces.

There is an array of suitable building materials for hardscapes. Keep in mind the building materials used throughout the home, and repeat the use of these materials in the landscape. When possible, use porous hardscape materials to

Utah House, Kaysville, UT. Photo and landscape design: Anne Spranger

Proper selection of perennial flowers is the key to achieving year-round show of color.

help keep landscape water runoff on-site. When porous materials are not used, redirect runoff to an area of the yard that can use the extra water. A well-planned water-efficient landscape usually has several hardscape features to provide interest and variety.

Do not make the mistake of substituting hardscapes for plants.

Landscape Design Principles

Unity in Design

A successful water-efficient landscape has year-round color and interest, functions well for the client, and of course, conserves water. Creating a unified or cohesive landscape is the ultimate goal in terms of design composition. A unified landscape has continuity, or a feeling of oneness. More simply stated, unity means that all parts of the landscape design go together; they fit. A unified

Utah House, Kaysville, UT. Photo and landscape design: Anne Spranger

Figure 1-14. A properly constructed arbor provides shade during the hottest parts of the day and can be used to support tall or vining plants. This arbor at the Utah House was designed with sustainability in mind and is constructed from recycled wood.

water-efficient landscape revolves around a set of aesthetic guidelines or design principles that help create a visually pleasing design solution. A unified landscape can be designed successfully by carefully combining the following design principles: simplicity, variety, balance, scale, emphasis, sequence, and interconnection. These design principles are applied to the physical properties of plants (form, color, texture) to achieve a unified and cohesive design.

Simplicity

Simplicity does not mean a design will be boring or dull. Simple lines and forms create appeal that is subtle, while allowing for proper focus of attention. The difficulty is in knowing what to eliminate while still creating an interesting design. Repetition is the most important factor when creating simplicity in design. Repetition refers to repeated use of form, color, or texture, as well as to repeated use of specific plants. Repetition can also apply to hardscape and home building materials. Too much repetition or simplicity can create monotony; but when used successfully, it can lead to rhythm, focus, or emphasis. Variety can be used to control simplicity and create interest for the viewer.

Variety

Variety adds interest and allows the designer to control the feeling of a landscape. By using a variety of forms, colors, textures, and plants, the designer can avoid a boring landscape, and instead create a landscape that keeps the viewer interested. Simplicity and variety must be carefully balanced. While too much repetition can be monotonous, too much variety can result in chaos. Variety in moderation is a sensible approach.

Balance

Although we may not realize it, we humans look for balance in everything we view. *Symmetrical balance* is achieved when one side of a design is a mirror image of the other side. A formal water-efficient design should have symmetrical balance. *Asymmetrical balance*, on the other hand, is marked by dissimilar elements of equal weight on each side of an axis. Asymmetrical balance should be used when designing an informal water-efficient landscape. The client's preferred landscape style as well as the architectural style of buildings on the site should influence the type of balance used in the landscape.

Scale

Trees and shrubs at mature size should be in proportion to the home.

Relative scale or proportion refers to the size of an object in relation to its surroundings. People tend to relate the size of an object to their own form. For example, a water-efficient garden designed especially for children should be designed differently in terms of scale than a garden designed primarily for adults. The designer can evoke emotions by controlling the proportionate scale of landscape features. The home will determine the proportion of the landscape. Trees and shrubs at mature size should be in proportion to the home. For example, an 80-foot-tall tree may look too large and out of scale next to a one-story ranch-style home. The relative scale of plantings, shade structures, fences, decks, and patios when compared to the human form should be considered.

Emphasis

Emphasis is a method used used by landscape designers to draw attention to important features in the landscape. Less important features then take on a secondary role. The primary entrance to a home or building is an area where attention is often desired. Emphasis can be placed through the use of plants that offer

Utah Botanical Center Welcome Pavilion, Kaysville, UT. Photo: Anne Spranger. Landscape design: Darrel Morrison, William Varga, and Anne Spranger

Figure 1-15. This informal naturalistic landscape includes repeating colors and hardscape elements that draw the viewer's eye through and around the landscape, providing a sense of organization.

outstanding form, color, or texture. Accent plants have especially strong features, and can be used to emphasize an area.

Sequence and Rhythm

Sequence and rhythm are achieved when the elements of a design create a feeling of motion, which leads the viewer's eye in an organized manner through the landscape. A gradual and subtle repetition of form, color, or texture can create rhythm in the landscape. Sequence and rhythm reduce confusion in the design.

Logan, UT. Photo and landscape design: Anne Spranger

Figure 1-16. Before (above) and after (below) installing a water-efficient landscape. Properly designed water-efficient landscapes not only conserve water, but they soften the edges and glare of building materials and create an inviting atmosphere for viewers and visitors to the home.

Interconnection

A design that lacks unity will look fragmented and disconnected. Interconnection is a method of joining elements of the design composition together, providing unity in the landscape. One way this can be accomplished is through the use of a ground cover that connects two separate planting areas together. Hardscape elements such as a fence, wall, patio, or deck can also be used to connect elements of the design composition together. Interconnection will help the designer create a cohesive and unified water-efficient landscape.

Planning and Design Summary

The first step to a successful water-efficient landscape is to have a plan. A well-thought-out plan will not only help to conserve water; it will also save you time, energy, and money. Water-efficient landscape designs can be formal, informal, or naturalistic; they can be filled with a variety of colors, textures, and forms that provide interest in every season. The principles of design that apply to water-efficient landscapes are not different from those of conventional landscapes. The focus of water-efficient landscape design, however, should be on using and directing landscape water such that water loss due to runoff or evaporation is minimal.

Client Questionnaire Checklist

CLIENT AND FAMILY INFORMATION

Name:____

Address:____

Contact Information:________________________________

Children:_______________________________________

PROGRAM DEVELOPMENT / LANDSCAPE GOALS AND OBJECTIVES

WATER USE

- ❑ Very Low
- ❑ Low
- ❑ Moderate

FUNCTIONAL NEEDS

- ❑ Privacy
- ❑ Erosion control
- ❑ Provide shade
- ❑ Children safety
- ❑ Other:______________________________________

AESTHETICS

- ❑ Curb appeal
- ❑ Beauty
- ❑ Colorful
- ❑ Year-round interest
- ❑ Fragrance
- ❑ Other:______________________________________

DESIGN STYLE

PREFERRED LANDSCAPE DESIGN STYLE

- ❑ Formal
- ❑ Informal
- ❑ Naturalistic
- ❑ Other:______________________________________

LANDSCAPE USE

ENTERTAINING

- ❑ Outdoor dining
- ❑ Furniture needs:____________________
- ❑ Cooking / Barbecue
- ❑ Number of people:____________________
- ❑ Formal
- ❑ Informal
- ❑ Covered space
- ❑ Other:____________________

CHILDREN'S PLAY

- ❑ Number and ages of children:____________________
- ❑ Creative play
- ❑ Play equipment
- ❑ Lawn
- ❑ Other:____________________

SPECIALTY GARDENS

- ❑ Kitchen garden: Fruits / Vegetables / Herbs / Nuts
- ❑ Fragrance garden
- ❑ Wildlife garden
- ❑ Children's garden
- ❑ Other:____________________

SPECIAL FEATURES

- ❑ Water feature
- ❑ Sculpture
- ❑ Pool
- ❑ Hot tub
- ❑ Fire pit
- ❑ Trellis
- ❑ Shade structure
- ❑ Arbor
- ❑ Other:____________________

SERVICE NEEDS

- ❑ Storage
- ❑ Garden tools

- ❑ 4-wheelers
- ❑ RV
- ❑ Boat
- ❑ Composting
- ❑ Recycling / Garbage
- ❑ Other:______________________________

*OTHER:*______________________________

SPECIAL CONSIDERATIONS

LANDSCAPE LIGHTING

- ❑ Function:______________________________
- ❑ Aesthetics

SAFETY

- ❑ Children
- ❑ Pets

FENCING

- ❑ Material:______________________________
- ❑ Style:______________________________
- ❑ Height:______________________________

NOISE

- ❑ Source:______________________________
- ❑ Location:______________________________
- ❑ Time of day:______________________________

CIRCULATION AND HARDSCAPE ELEMENTS

DRIVEWAY

- ❑ Existing condition:______________________________
- ❑ Parking needs:______________________________
- ❑ Material selection:______________________________

ENTRY WALK

- ❑ Existing condition:______________________________
- ❑ Materials:______________________________
- ❑ Size:______________________________

SECONDARY PATHS

- ❑ Needs:____________________
- ❑ Materials:____________________
- ❑ Function:____________________

DECK / PATIO / COURTYARD

Number of People / Size:____________________

Use:____________________

SPECIAL NEEDS

- ❑ ADA Compliance
- ❑ Other:____________________

IRRIGATION

IRRIGATION SOURCE

- ❑ Culinary
- ❑ Secondary (canal or well)
- ❑ Reclaimed (recycled)

NEED FOR AUTOMATIC IRRIGATION SYSTEM

- ❑ Yes
- ❑ No

*BUDGETED ALLOTMENT:*____________________

DESIRE FOR WATER-HARVESTING

- ❑ Roof collection system
- ❑ Rain barrel
- ❑ Underground water-collecting cistern

PLANT SELECTION

FAVORITE PLANTS:____________________

PLANT DISLIKES:____________________

PREFERRED COLORS:____________________

PLANT ALLERGIES:____________________

MAINTENANCE

DESIRED MAINTENANCE LEVEL

- ❑ High
- ❑ Medium
- ❑ Low

ACCEPTABLE MAINTENANCE PRACTICES

- ❑ Mowing
- ❑ Pruning
- ❑ Deadheading
- ❑ Weeding

BUDGET

ALLOTTED LANDSCAPE BUDGET

- ❑ Conservative
- ❑ Moderate
- ❑ Sky is the Limit

TYPICAL ELEMENTS THAT MAY BE INCLUDED IN THE BUDGET

- ❑ Landscape design fee
- ❑ Earthwork / Grading
- ❑ Topsoil
- ❑ Irrigation: design, material, and installation
- ❑ Hardscapes: design, material and installation
- ❑ Landscape Structures: design, material, and installation
- ❑ Plant material and installation
- ❑ Mulch material and installation

Principle 2

Soil Preparation

Grant E. Cardon, Ron Patterson, and Rick Heflebower

Introduction

Fundamental to water-efficient management in landscapes and gardens is an understanding of how soils interact with water and how various soil properties affect the infiltration, movement, and retention of water. In this section, we briefly introduce soils and soil properties, emphasizing those important to water relations. We also discuss manipulation of those properties to promote or facilitate good plant/water relations.

Any given soil volume comprises four main components: unconsolidated mineral matter, organic matter, air, and water. The typical relative proportions of the components are shown in Figure 2-1.

Soil Mineral Matter

The mineral matter of the soil refers to the content of sand, silt, and clay. The proportions of these particle types give a soil its particular texture. Sand, silt, and clay are size classes of mineral particles, each imparting a characteristic influence on the soil's physical, chemical, and hydraulic properties. The relative particle sizes that make up each class are shown in Figure 2-2.

Individual sand particles are visible to the naked eye and provide the coarse, gritty feel of a soil. Silt particles are at the lower margin of the size class of sand

and are nearly one thousand times smaller than the largest sand particles. Both sand and silt particles are roughly spherical to blocky in shape.

Clay particles, on the other hand, are very small, platelike particles crystallized from the weathering products of the soil's primary minerals, or washed or blown onto a soil from other areas. Clay particles are from one thousand to many thousand times smaller than sand particles. The platy shape of these clay particles means that they have an enormous amount of surface area per unit volume, which provides habitat for soil microbes, as well as reactive sites for chemical reactions and water adsorption. One gram of clay can have as much as 1,000 square meters of reactive surface area. Therefore, clay exerts some of the strongest influence on a soil's capacity for water retention.

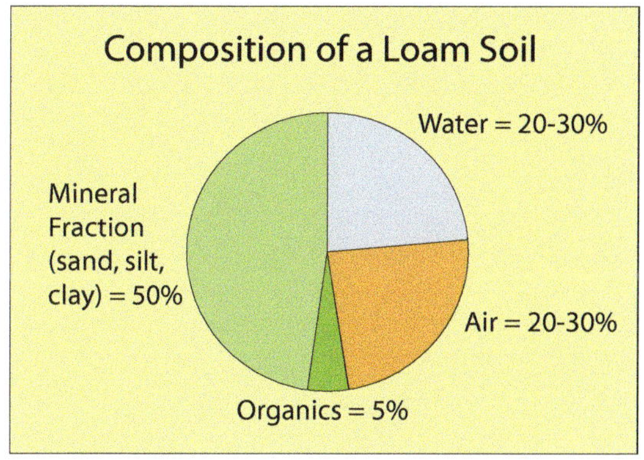

Figure 2-1. Typical ranges for the four components in a volume of intermountain western soil.

The relative proportion of sand, silt, and clay in the mineral matter of a soil is termed its *soil texture*. Soils having similar characteristics based on their texture are grouped together and named for convenience as shown in Figure 2-3. This is known as the soil textural triangle. Textural group names provide a ready reference to the proportion of sand, silt, and clay in a soil, as well as its suite of physical, chemical, and hydraulic properties.

Figure 2-2. Relative size classes and textural character of soil particles.

For instance, if one refers to a sandy loam and a silty clay, it is immediately understood that the two soils are widely divergent in their character because of their distance from one another on the soil textural triangle. The sandy loam soil has a highly gritty feel, larger particles and therefore larger pores, and much less reactive surface area than the silty clay. On the other hand, the silty clay is fairly sticky when wet and has a high water retention capacity, but will not allow for easy water infiltration due its much smaller pores compared to the sandy loam. Note the strong influence of clay content on the characteristics of soils. Above about 40% clay content (the upper third of the textural triangle), soils overwhelmingly take on the character of the clay particles and form a large single group.

Soils in the loam category and in the immediately adjacent loam-modified categories have the optimal combination of large and small soil particles, large and small pores, and an ideal amount of reactive surface area. These loamy soils tend to drain well, but they can hold sufficient water and nutrients in their pore spaces and on particle surfaces to provide for plant growth and optimal water relations. Soils at the extremes of texture are generally overwhelmed by the individual characteristics of their primary particle size and either do not hold much water

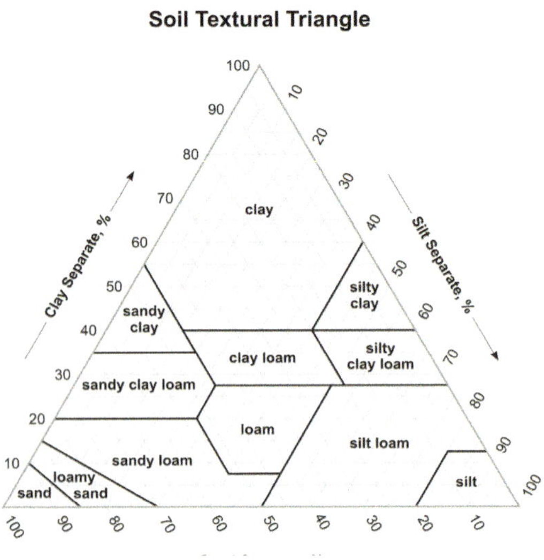

Figure 2-3. The soil textural triangle.

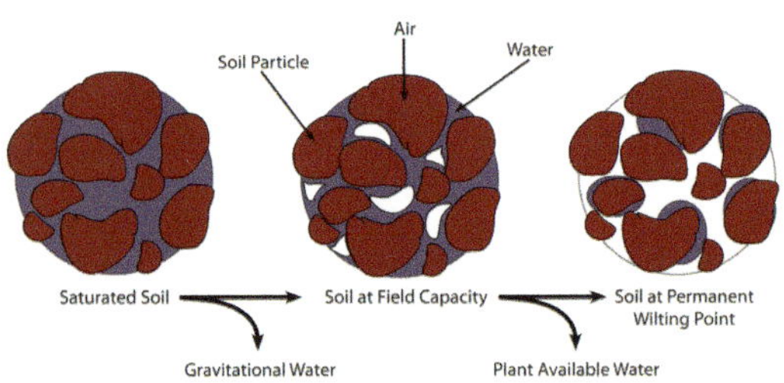

Figure 2-4. Schematic diagram of water held in soil.

(sands), resist the infiltration of air and water (clay), or do not hold together well and are easily eroded (silts).

Plants use water that is stored in a soil's pores and adsorbed to soil surfaces (Figure 2-4). However, as a soil dries out from plant water extraction and evaporation (together called evapotranspiration, or ET), it becomes increasingly difficult for the plant to remove water from the soil. At a certain point, a given soil will no longer yield usable quantities of water to the plant, and the plant will wilt. If this condition persists, the plant will permanently wilt and die.

Because of their large pores and low surface area, sands cannot store much water for plant use. What little water is stored, however, is easily released to the plant. Clays, on the other hand, because of their large surface area and small pores, can store large quantities of water. As water is removed through the action of ET, however, the stored water becomes harder and harder to remove, as smaller and smaller pores remain filled, and water films on soil surfaces become quite thin. Clays, therefore, hold more water than loamy soils, but do not release it as easily for plant use. Hence, loamy soils tend to have the largest reservoir of *plant-available water* (the portion of stored water that can be used by the plant). Average values of plant-available water in a given soil texture are provided in Table 2-1 for reference.

Table 2-1. Average values of plant-available water for soils of different texture.

Soil Textural Class	Inches of plant-available water per foot of soil depth
Coarse sand	0.60 – 0.80
Fine sand	0.80 – 1.00
Loamy sand	1.10 – 1.20
Sandy loam	1.25 – 1.40
Fine sandy loam	1.50 – 2.00
Loam	2.20 – 2.50
Silty loam	2.00 – 2.50
Silty clay loam	1.80 – 2.00
Silty clay	1.50 – 1.70
Clay	1.30 – 1.50

Soil Organic Matter

In intermountain western soils, organic matter content varies from 0.5% to 2.0% by weight depending on the aridity (dryness) of the location and the plants growing at the site. One might argue that the proportion of organic matter is so small in intermountain western soils as to be insignificant to their character. That would be true except for the very high water-holding capacity and surface reactivity of organic matter.

Organic matter can hold as much as 20 times (or 2,000%) its weight in water. Soil minerals, on the other hand, hold between 15% and 20% of their weight in water. Therefore, very small quantities of organic matter can impart significant characteristics to a soil. In fact, *the single most effective soil amendment to enhance water-holding capacity, water release, soil infiltration, and a number of other important soil properties critical to water relations, is organic matter* (see Table 2-2 for common soil organic amendments in the Intermountain West).

Table 2-2. Soil organic amendments available in the Intermountain West.

ORGANIC AMENDMENT	ADVANTAGES	DISADVANTAGES	USE
Wood residues (ground bark, wood shavings, sawdust)	Slow to decompose	May tie up nitrogen (N) in the soil if not composted	Incorporate 1 lb. N per 100 lb. material added
Sphagnum peat	Acidifies the soil; improves soil drainage	Expensive	Incorporate thoroughly into soil
Peat moss	Acidifies the soil; improves water-holding capacity of sandy soils	Quality varies; not effective in clay soils	Incorporate thoroughly into soil
Steer manure	High in nutrients	May be high in salts; may contain weed seeds	Use only well-composted manure; use in small quantities
Green waste (grass clippings, chopped leaves)	Readily available; makes use of recycled materials	Decomposes rapidly	Incorporate thoroughly into soil
Soil Pep (composted bark product)	Improves soil drainage and structure	Low in nutrients; may need to add N	Watch plants for signs of deficiency

Adapted from Utah State University Extension publication, "Solutions to Soil Problems: V. Low Organic Matter." http://extension.usu.edu/files/publications/publication/AG_Soils_2003-05.pdf.

On the other hand, the addition of sufficient mineral matter to correct a soil texture imbalance is scale-prohibitive. The top foot of a mineral soil over a 500-square-foot garden area, for example, weighs approximately 40,000 pounds, or 20 tons. To change the sand or silt content by even 1% by weight would require 400 pounds of material. To go from a sandy loam to a loam through the addition of clay would require about a 10% increase, or 4,000 pounds of clay—a highly impractical addition.

The primary focus of any effort to improve water relations in soils is to promote the development of soil structure.

Small additions of organic matter, however, effect tremendous changes in soil properties with respect to water relations. Over time, organic matter content can be built up in soil to 1%–2% over existing conditions, with annual additions. Annual additions are critical, however, because soil microbes, worms, and other soil *biota* (life) cause the breakdown and removal of organic matter from the soil system.

Other Soil Properties

In a healthy, functioning soil system, all the soil components (mineral and organic solids, air, and water), the microbes and other soil biota (worms, ants, beetles, bacteria, fungi), and growing plants, work together to transform the otherwise inert unconsolidated mineral matter at the earth's surface into a living, breathing tissue.

The primary focus of any effort to improve water relations in soils is to promote the development of soil structure. Soil structure is the formation of aggregates, each containing the various components of the soil system that act as an individual unit. As structural aggregates form in soils, infiltration, aeration, and water retention are increased, allowing the system to function more efficiently, and providing improved conditions for further soil development.

The primary method of promoting soil structure is the regular addition of organic matter.

Examples of soil structural aggregates are shown in Figures 2-5 and 2-6. Soil structure is a telltale sign of good and/or improving soil quality. Aggregates can form only as soil mineral matter, organic matter, and the action of plants and soil biota combine. Soils with adequate structure are easy to till and work, and are *friable* (crumbly) enough for the proper growth and elongation of plant roots. If a soil begins to lose structure (aggregates begin to break down and leave behind individual soil mineral particles), it is a sign that improvements are needed.

The primary method of promoting soil structure is the regular addition of organic matter. Well-composted materials are best because much of the initial breakdown of the material has taken place during the composting process, and the material can begin to build soil structure and improve water relations fairly quickly.

Additions of uncomposted organic matter also are desirable, but may require some special considerations. Lawn clippings and other green yard materials (including incorporated cover crops) contain sufficient nitrogen to support their rapid breakdown when introduced fresh into the soil. Brown or woody plant materials (dry leaves, sawdust, chipped limbs, and branches) are low in nitrogen and will cause a decrease in soil nitrogen if directly incorporated into the soil. These materials should be composted prior to soil incorporation, or added to the soil along with additional nitrogen to provide for their breakdown. The recommendation when incorporating uncomposted woody plant materials into the landscape is to add one pound of nitrogen for every 100 pounds of material added. Raw manure can be added as well, but can be high in salt content and is often a source of weed seed. It should be tested for salinity prior to application in the event that salt leaching is needed.

The recommendation when incorporating uncomposted woody plant materials into the landscape is to add one pound of nitrogen for every 100 pounds of material added.

Photo: Grant Cardon

Figure 2-5. Soil granular structure

Photo: Grant Cardon

Figure 2-6. Soil structure development as a result of organic matter additions and root action.

Topsoil Guidelines

Every soil can be managed to improve soil/water/plant relations, particularly through the addition of organic matter and the promotion of soil biotic activity. Soils at the extremes of texture (coarse sandy soils, high-clay-content soils), however , may present challenges too laborious or costly to overcome on a practical level. In those cases, it may be more useful to remove the existing soil and replace it with higher quality topsoil.

The preceding primer on soil properties related to optimum soil/water/plant relations will be useful in selecting or purchasing topsoil in such cases. There are numerous sources of soil materials in the Intermountain West. Take care to ensure that the material you are importing has, in fact, the desirable characteristics discussed. An excellent guide to optimum topsoil conditions is provided in Table 2-3. Inexpensive soil tests (such as those offered through the Utah State University Analytical Laboratory, http://www.usual.usu.edu) should be sought prior to the purchase and spreading of any topsoil.

Special Soil Preparation Guidelines for Growth of Intermountain Western Native Plants

Most intermountain western native plants (IWNP) are accustomed to arid and semiarid conditions, and are popular additions in many water-efficient landscaping plans. Although such plants are desirable for water conservation, they present several challenges during their establishment in landscapes. One of the most important factors affecting their establishment is the proper preparation of the soil prior to planting.

Much of the challenge in establishing IWNP is the competition that occurs when non-native and weedy species are grown in close proximity. Reduction of competition for water and nutrients is, therefore, a key to early establishment of IWNP. Where there is a high level of competition from weeds in an area that is to be planted with IWNP species, consider cleaning up the existing weed seed bank. To accomplish this, disturb the top six inches of soil periodically to promote weed seed germination. Follow this with careful weed control by removing or spraying the germinated weeds before they go to seed. This process can be repeated several times in a growing season, and may require up to a year to reduce the viable but dormant weed seed supply in the soil profile.

In addition to preconditioning the soil by reducing weed seed populations, consider the use of pre-emergent, short-term herbicides that can be applied after transplanting IWNP species. These herbicides attack newly emerging weeds and provide a competitive advantage to existing vegetation. Be careful to use short-lived products. Long-term activity of herbicides in soils will reduce the health of existing vegetation. Preconditioning the soil with a sterilant-type herbicide (broad spectrum, multiyear activity) is never advised in an area that is to be planted. These chemicals will continue to have residual action on all plants growing in the soil, making it very difficult or impossible to establish vegetative cover.

Intermountain western native plants are much better suited to their native soil conditions. Soils that are too high in organic matter content (for example,

Category	Soluble salts (dS/m or mmhos/cm	pH	Sand (%)	Silt (%)	Clay (%)	Texture class*	Organic matter (%)	% Coarse fragments (> 2 mm in diameter)**	Sodium Adsorption Ratio (SAR)*
Ideal	< 2	5.5 to 7.5	< 70	< 70	< 30	L, SiL	≥ 2.0	≤ 2	< 3 for any texture
Acceptable	< 4	5.0 to 8.2	< 70	< 70	< 30	SCL, SL, CL, SiCL	≥ 1.0	2.1 to 5.0	3 to 7 (SiL, SiCL, CL) 3 to 10 (SCL, SL, L)
Unacceptable	> 4	< 5.0 or > 8.2	> 70	> 70	> 30	LS, SC, SiC, S, Si, C	< 1.0	> 5.0	> 10 for any texture

*L = loam; SiL = silt loam; SCL = sandy clay loam; SL = sandy loam; CL = clay loam; SiCL = silty clay loam; LS = loamy sand; SC = sandy clay; SiC = silty clay; S = sand; Si = silt; C = clay.
**No fragments should be larger than 1 ½ inch in diameter. (adapted from Koenig and Isaman, 2002)

Table 2-3. Topsoil quality guidelines for the Intermountain West. The units for soluble salts are deciSiemens per meter (dS/m) or millimhos per centimeter (mmhos/cm).

soils heavily amended with peat) tend to promote non-native species over many native species. An attempt should be made to match native soil conditions by ensuring that the soil contains no more than 2% organic matter, has good granular structure (Figures 2-5 and 2-6), and is maintained at the minimum possible moisture level for the species (see "Principle 3: Appropriate Plant Selection" for drought-tolerance ratings of many IWNP species). Preparing soils two to three months in advance to allow for soil settling, organic matter breakdown, and soil structure development and moisture regulation, is helpful in promoting IWNP establishment.

With respect to soil moisture management, *hydrozoning* (also called "water-use zoning") is critical to the optimum performance of a landscape containing IWNP species (see "Principle 6: Irrigation Planning"). Zones containing IWNP should be watered so that no more than the natural environmental mean precipitation is applied during the growing season. For many IWNP, minimal watering (except for the first growing season after planting) will promote optimal establishment and long-term performance. Plants can be supplemented with irrigation water during periods of extended drought or particularly hot, dry periods, but care should be taken to make sure that the total applied water is reflective of the natural conditions to which the species is adapted.

Soil fertility management is another important element in the establishment and optimum performance of IWNP. Care should be taken to restrict the broadcasting of highly soluble nutrient sources (such as most chemical fertilizers). Easily soluble, highly available nutrients promote the growth of non-native and weedy species. Instead, consider building soil nutrient content through regular additions of well-composted, low-nutrient organic materials, and the promotion of soil biota. Localized additions of starter fertilizers (such as 5-10-10) within the root zone of IWNP can help minimize transplant shock and promote more rapid establishment. Broadcast applications of fertilizers, however, promote growth of weedy or non-native species.

Some biostimulant products (humic and fulvic acids, humates, kelp, and other marine bioproducts) have shown some benefit for IWNP establishment. These products can tip the balance in favor of the native species over the non-native and weedy species. Although biostimulant amendments have shown some promise, bacterial and fungal inoculants, especially mycorrhizal inoculants, have not. Mycorrhizae are so specific to individual species that it is not practical to try to inoculate with a general mix of these soil biota. It is better to *promote* the growth of specific mycorrhizae already present in the soil and roots of plants by building soil conditions optimum for their growth. In other words, build the proper environment and the natural system will do its job.

Long-term soil management is as important to IWNP performance as the initial soil preparation. Care should be taken to minimize soil disturbance. This reduces weed seed cycling from deeper depths and suppresses weed seed germination and growth, reduces the potential for soil structure breakdown, and minimizes the potential for the soil compaction that can occur during tillage or via implement and foot traffic on the soil surface.

Additionally, long-term soil organic matter levels are best maintained with higher- carbon-content materials. Fresh leaf and lawn waste high in nitrogen should not be used. Well-composted woody materials, animal wastes, and dry leaf waste should be used. Mulching with these materials is an excellent way of protecting the soil surface, suppressing weed growth, minimizing evaporation, and cycling organic matter into the soil (see "Principle 5: Use of Mulch").

When applying animal wastes (like manure-based composts) for organic matter additions, confirm that the composting process has reached an internal temperature of >140°F to ensure weed seed destruction, and that enough leaching has occurred to reduce the salinity of the material (measured as electrical conductivity or EC) to <10 dS/m. Add no more than one-half to three-quarters of an inch of material annually (see "References and Additional Resources" for further information on using composts in landscapes).

Principle 3

Appropriate Plant Selection

Amy Croft and Heidi Kratsch

It is a misconception that water-efficient landscapes must be designed using only plants native to the region. Some plants native to the Intermountain West are particularly adapted to the higher-elevation, dry, and often harsh conditions of the region. But many plants native to other parts of the country and the world with similarly harsh climates perform just as well. The suggested plant list at the end of this chapter includes both native and non-native plant species and cultivars that have demonstrated outstanding performance in managed landscapes in our climate. It is important to understand the characteristics that have made these plants drought-adapted and, therefore, unique in the ways they perform and the treatments they require under managed conditions. For this reason, the first part of this chapter is devoted to an explanation of how plants have adapted to arid environments.

Water-Efficient Plant Adaptations

Arid and semiarid climates are characterized by the scarce and variable nature of their water resources. Therefore, plant life in such climates is conditioned by available moisture. Plants native to arid and semiarid climates have adapted with characteristics that enable them to withstand drought situations. Adaptations include various leaf formations and angles to the sun, leaf colors, and different types of root systems. Some plants utilize more than one of these traits. All of these characteristics help a plant tap into or preserve the limited water necessary for its survival.

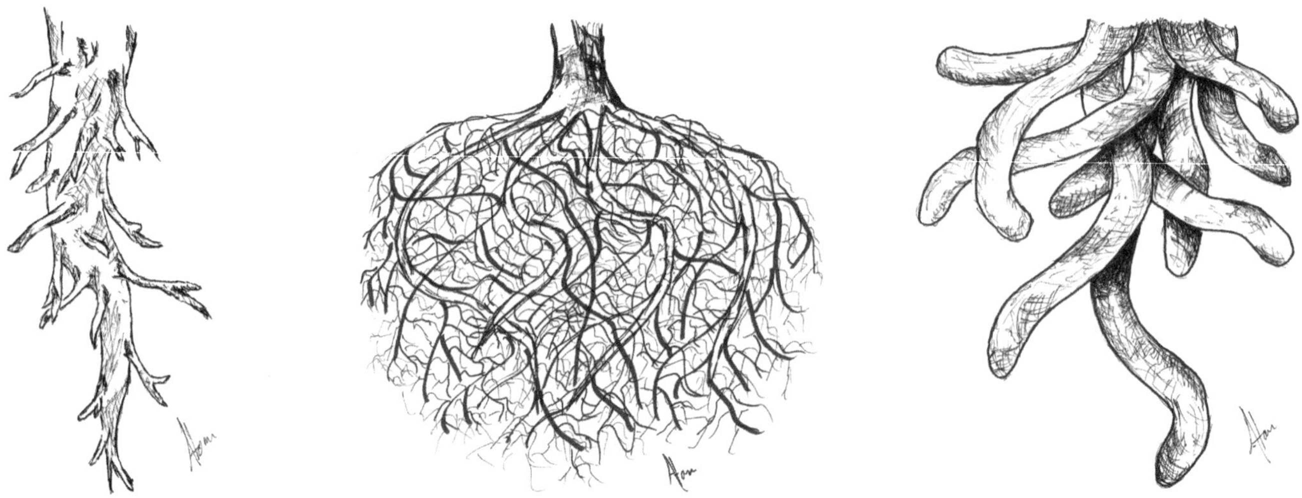

illustration: Adam Richards

Figure 3-1. a. Tap root. b. Sinuous roots. c. Fleshy roots.

Most plants adapted to dry climates have extensive root systems that allow the plants to take advantage of all available water resources. Some plants have deep taproots that are able to draw from permanent ground water (Figure 3-1a). Some have sinuous roots that are able to penetrate into rocky cracks and crevices or follow narrow seams of soil to gain access to water (Figure 3-1b). Some have fleshy roots that can store water during periods of drought (Figure 3-1c). Others have a shallow but widespread root system to take advantage of light summer rains. Some plants exhibit a combination of root systems. For example, yucca has a deep taproot but also has a shallow root system, thus enabling it to grow with no supplemental irrigation. Many plants native to arid and semiarid landscapes have root systems that are much more extensive than their shoots.

An extensive root system ensures plant survival during periods of drought, and such plants are ideal for a water-efficient garden. Because drought-tolerant plants rely so much on their root systems for survival, it is critical that their root systems are healthy. Healthy roots take in water and soil nutrients essential to plant health. Thus, a successful water-efficient landscape will be one that accommodates the root systems of xeric plants. This means that the soil bed must be prepared in such a way as to welcome their growth habit. Most xeric plants prefer well-drained soil. Excess moisture around a plant adapted for drought conditions can be detrimental. If the soil does not drain well, it should be amended to create more pore space, or special attention should be given to the amount of water applied to a xeric plant growing in such soil (see "Principle 2: Soil Preparation" for more information).

The most common cause of plant death in a water-efficient landscape is overwatering.

The most common cause of plant death in a water-efficient landscape is overwatering. Therefore, plants should be watered frequently only until root systems are established. Plants are not fully established until vigorous twig or top growth returns. For herbaceous plants, this may take several months. For trees and shrubs, this may take one to two years. Trees dug from the field take longer because most of their roots are left behind. Once plants are established, water should be applied only when necessary. In general, low-water-use plants should be watered thoroughly, then left unwatered until the soil is dry or until they show water stress. Irrigation needs will vary within each hydrozone, so this could be as often as once a week or as little as once a month (see "Principle 6: Efficient Irrigation" for specific watering needs based on hydrozone). Lawns and annual plants often require more frequent irrigation than trees and shrubs because of their relatively shallow root systems (see "Principle 4: Practical Turfgrass Areas" for more information on lawn care). Generally, the deeper the root system, the longer the interval should be between irrigation events. The rate of water application for a water-efficient landscape should be slow enough for penetration without runoff.

In addition to an extensive root system, many drought-adapted plants have modified leaves that help reduce the amount of water that is transpired from the leaf surface. Leaf adaptations vary and include leaf hairs, curled leaves, waxy leaves, succulent leaves, blue-tinted leaves, and small leaves. Leaf hairs reduce wind velocity over leaf pores and also provide a small amount of shade for the leaf surface, minimizing water loss; examples include the sages (*Artemisia* spp.) and the salvias (*Salvia* spp.) (Figure 3-2a). The leaf hairs on these plants also give the leaves a whitish cast, which reflects hot sunlight. The leaves of some plants are curled or rolled, thus reducing surface exposure to hot sun and wind; an example is the native shrub, curl-leaf mountain mahogany (*Cercocarpus ledifolius*). Other plant leaves have a waxy coating that directly limits the amount of water lost through surface

Photo: Larry Rupp

Photo: USU Herbarium

Photo: Roger Kjelgren

Photo: Roger Kjelgren

Figure 3-2. Examples of plants with leaf adaptations that enhance their drought tolerance. (a) *Salvia dorrii* – hairy leaves; (b) *Opuntia polyacantha* – spines; (c) *Mahonia fremontii* – blue-green leaves; (d) *Purshia mexicana* – small leaves.

evaporation. Some leaves, such as those of agave and yucca, are succulent and act as water storage devices. Cactus leaves are much reduced (spines) (Figure 3-2b); their fleshy stems take advantage of a waxy coating to reduce transpiration water loss. Many drought-tolerant plants have blue-green or gray leaves that reflect hot sunlight; an example is Utah holly (*Mahonia fremontii*) (Figure 3-2c). Many others have smaller than average leaves, thus reducing the transpiring surface area that is exposed to sun and wind; an example is the native shrub, cliffrose (*Purshia mexicana*) (Figure 3-2d). Look for plants with these notable leaf characteristics when choosing what to plant in a water-efficient landscape. They indicate a likely capacity to withstand sun, wind, heat, and drought. The plant list at the end of this chapter outlines plant species and cultivars with proven performance in intermountain western landscapes.

Planting Techniques

Although low-water-use plants are adapted to drought, their root systems must be established in order to demonstrate this characteristic. Care must be taken to minimize the stress of transplanting in order to increase the probability of successful establishment. Proper timing, temperature, moisture, soil environment, and handling are essential. The optimal time to plant is when transpiration rates are low. This can be accomplished when a plant is dormant, or in the spring or fall when plant growth is minimal but root regeneration is high. There must also be sufficient soil moisture for new roots to develop. Although planting can be done in the summer, it requires careful monitoring and more water than when conditions are more favorable in the spring or fall. When planting, it is recommended that planting holes be one to two inches shallower and at least two to three times wider than the root ball. This creates an aerated zone in which roots can become established. Soil should be backfilled into the planting hole such that the plant is sitting at or slightly above its original growing level. It is also helpful to build a collar or moat around the planting hole to help retain water during establishment.

The key to a successful water-efficient garden or landscape is to plant the right plants in the right place.

The key to a successful water-efficient garden or landscape is to plant the right plants in the right place. A water-efficient garden or landscape takes advantage of a concept known as *hydrozoning*. Hydrozoning is placing plants with similar water needs together. Therefore, there might be a high-water-use zone in the landscape as well as one that is not watered at all. In order to take advantage of the water and slopes available in a landscape, it is wise to plant high-water-use plants in low-lying drainages, near downspouts, in the shade of other plants, and where it is easy to water. Dry, sunny areas and those that are more difficult to water are ideal for low-water-use plants. (More detailed information about planning landscape hydrozones can be found in "Principle 1: Planning and Design.") Water conservation can be achieved through proper water application and selection of native landscape plants or non-natives that are adapted to the dry environment of the Intermountain West.

Helpful Resources for Selecting Water-Efficient Plants:

Web Sites:

http://www.waterwiseplants.utah.gov/
http://theutahhouse.org
http://utahbotanicalcenter.org/
http://www.centralutahgardens.org/
http://www.utahschoice.org/~inpga/natives/wheretobuy
(This site provides information on plant suppliers.)

Demonstration Gardens:

Utah Botanical Center. 700 South 150 East, Kaysville, Utah. (801) 593-8969 http://www.utahbotanicalcenter.org

Utah House. 920 South 50 West, Kaysville, Utah. (801) 544-3089 http://theutahhouse.org

Conservation Garden Park at Jordan Valley. 8215 South 1300 West, West Jordan, Utah. (801) 565-4300 http://www.conservationgardenpark.org

Central Utah Water Conservancy District. 355 West University Parkway, Orem, Utah. (801) 222-0123 http://www.centralutahgardens.org

Utah State University Greenville Research and Demonstration Gardens. 800 East 1800 North, North Logan, Utah http://www.hort.usu.edu/waterwisegardens

Red Butte Garden. 300 Wakara Way, Salt Lake City, Utah. (801) 581-IRIS (4747) http://www.redbuttegarden.org

Water-Efficient Landscaping Plant List

Trees, conifer

LATIN NAME	COMMON NAME	WATER-USE RATING	MAXIMUM HEIGHT (FT)	MAXIMUM SPREAD (FT)	FLOWER COLOR	BLOOM PERIOD	SUN/SHADE	HARDINESS ZONE
Abies concolor	White fir	Low	100	20	N/A	N/A	Full/part	3–7
Calocedrus decurrens	Incense-cedar	Low	90	15	N/A	N/A	Full/part	5–8
Cedrus atlantica 'Glauca'	Blue atlas cedar	Medium	60	30	N/A	N/A	Full/part	6–9
Cedrus libani	Cedar of Lebanon	Medium	60	50	N/A	N/A	Full sun	5–7
Cupressus arizonica	Arizona cypress	Low	40	20	N/A	N/A	Full sun	7–9
Juniperus osteosperma	Utah juniper	Very Low	20	10	N/A	N/A	Full sun	3–7
Juniperus scopulorum	Rocky Mountain juniper	Low	20	15	N/A	N/A	Full sun	3–7
Picea pungens	Blue spruce	Medium	80	30	N/A	N/A	Full/part	2–7
Pinus edulis	Pinyon pine	Low	35	15	N/A	N/A	Full sun	4–8
Pinus flexilis	Limber pine	Medium	35	30	N/A	N/A	Full sun	2–7
Pinus flexilis 'Vanderwolf's Pyramid'	Vanderwolf's Pyramid limber pine	Medium	30	15	N/A	N/A	Full sun	4–7
Pinus heldreichii var. leucodermis	Bosnian pine	Medium	60	40	N/A	N/A	Full sun	5–10
Pinus longaeva (aristata)	Bristlecone pine	Low	25	20	N/A	N/A	Full sun	2–6
Pinus monophylla	Singleleaf pinyon	Low	25	15	N/A	N/A	Full sun	5–9
Pinus nigra	Austrian pine	Low	60	30	N/A	N/A	Full sun	4–7
Pinus ponderosa	Ponderosa pine	Low	80	40	N/A	N/A	Full sun	3–7
Pinus sylvestris	Scots pine	Low	70	30	N/A	N/A	Full sun	2–8
Pseudotsuga menziesii	Douglas-fir	Medium	80	25	N/A	N/A	Full sun	2–5

Trees, deciduous

LATIN NAME	COMMON NAME	WATER-USE RATING	MAXIMUM HEIGHT (FT)	MAXIMUM SPREAD (FT)	FLOWER COLOR	BLOOM PERIOD	SUN/SHADE	HARDINESS ZONE
Acer campestre	Hedge maple	Medium	30	30	N/A	N/A	Full/part	4–8
Acer ginnala 'Flame'	Flame amur maple	Low	20	20	N/A	N/A	Full/part	2–8
Acer grandidentatum	Bigtooth maple	Low	30	25	N/A	N/A	Full/part	3–6
Acer tataricum	Tatarian maple	Low	25	20	N/A	N/A	Full sun	3–8
Acer truncatum	Shantung maple	Medium	25	20	N/A	N/A	Full/part	4–8
Amelanchier alnifolia	Saskatoon serviceberry	Medium	12	10	White	Early	Full/part	3–8
Amelanchier utahensis	Utah serviceberry	Low	12	10	White	Early	Full/part	3–8
Catalpa speciosa	Western catalpa	Medium	60	40	White	Mid	Full/part	4–9
Celtis occidentalis	Common hackberry	Medium	50	40	N/A	N/A	Full/part	2–9
Celtis reticulata	Netleaf hackberry	Low	30	20	N/A	N/A	Full/part	4–9
Chilopsis linearis	Desert willow	Low	30	20	Pink	Early–Mid	Full sun	6–10
x Chitalpa tashkentensis	Chitalpa	Low	30	30	Pink	Mid	Full sun	6–10
Corylus colurna	Turkish filbert	Medium	50	25	N/A	N/A	Full sun	4–8
Crataegus douglasii	Douglas hawthorn	Low	12	10	White	Early	Full sun	2–8
Forestiera neomexicana	Desert olive	Low	10	8	N/A	N/A	Full sun	5–9
Fraxinus anomala	Singleleaf ash	Low	12	12	N/A	N/A	Full sun	6–9
Ginkgo biloba	Maidenhair tree	Low	50	30	N/A	N/A	Full sun	3–9
Gymnocladus dioicus	Kentucky coffeetree	Low	60	40	Cream	Early	Full/part	3–8
Koelreuteria paniculata	Goldenraintree	Low	35	40	Yellow	Mid	Full sun	5–9
Maclura pomifera	Osage-orange	Low	40	40	N/A	N/A	Full sun	4–9
Malus 'Braeburn'	Braeburn apple	Medium	15	15	White	Early	Full sun	5–8
Prunus virginiana	Chokecherry	Medium	20	15	White	Early	Full sun	2–6
Quercus gambelii	Gambel oak	Low	30	15	N/A	N/A	Full sun	4–8
Quercus macrocarpa	Bur oak	Low	75	70	N/A	N/A	Full sun	2–8
Quercus muehlenbergii	Chinkapin oak	Low	50	50	N/A	N/A	Full/part	4–7
Robinia neomexicana	New Mexico locust	Low	10	6	Pink	Mid	Full sun	5–9
Syringa reticulata	Japanese tree lilac	Medium	30	20	White	Early–Mid	Full sun	3–8
Ulmus parvifolia	Lacebark elm	Low	50	50	N/A	N/A	Full/part	4–9
Zelkova serrata	Japanese zelkova	Low	60	60	N/A	N/A	Full sun	5–8

Shrubs, evergreen

LATIN NAME	COMMON NAME	WATER-USE RATING	MAXIMUM HEIGHT (FT)	MINIMUM HEIGHT (FT)	FLOWER COLOR	BLOOM PERIOD	SUN/ SHADE	HARDINESS ZONE
Arctostaphylos patula	Greenleaf manzanita	Low	6	1	Pink	Early–Late	Full sun	2–6
Artemisia filifolia	Sand sagebrush	Very Low	4	3	Yellow	Mid–Late	Full sun	4–8
Artemisia nova	Black sagebrush	Very Low	2	2	Yellow	Mid–Late	Full sun	3–6
Artemisia tridentata	Big sagebrush	Low	5	5	Yellow	Mid–Late	Full sun	3–6
Atriplex canescens	Fourwing saltbush	Very Low	4	6	N/A	N/A	Full sun	2–10
Atriplex confertifolia	Shadscale	Very Low	3	3	N/A	N/A	Full sun	3–6
Krascheninnikovia lanata	Winterfat	Very Low	2.5	2.5	White	Mid	Full sun	2–7
Cercocarpus intricatus	Littleleaf moun-tain mahogany	Low	5	4	Pink	Early	Full sun	4–7
Cercocarpus ledifolius	Curl-leaf moun-tain mahogany	Very Low	12	12	Pink	Mid	Full/part	3–8
Cotoneaster microphyllus var. *thymifolius*	Dwarf rockspray cotoneaster	Low	2	4	White	Early–Mid	Full sun	5–11
Ephedra viridis	Mormon tea	Very Low	4	4	N/A	N/A	Full sun	3–8
Juniperus squamata 'Blue Star'	Singleseed juniper	Low	3	4	N/A	N/A	Full sun	4–8
Mahonia fremontii	Utah holly	Very Low	5	5	Yellow	Early–Mid	Full sun	3–6
Pinus mugo 'Slowmound'	Slowmound mugo pine	Low	3	3	N/A	N/A	Full/part	2–7
Purshia mexicana	Cliffrose	Very Low	5	4	White	Early	Full sun	4–7
Purshia tridentata	Antelope bitterbrush	Low	8	6	Yellow	Early	Full sun	3–6
Shepherdia rotundifolia	Roundleaf buffaloberry	Very Low	6	6	Yellow	Early–Mid	Full sun	5–9

Shrubs, deciduous

LATIN NAME	COMMON NAME	WATER-USE RATING	MAXIMUM HEIGHT (FT)	MINIMUM HEIGHT (FT)	FLOWER COLOR	BLOOM PERIOD	SUN/ SHADE	HARDINESS ZONE
Amorpha canescens	Leadplant	Low	4	3	Purple	Mid–Late	Full sun	4–8
Caragana pygmaea	Pygmy peashrub	Medium	3	5	Yellow	Mid–Late	Full sun	3–7
Caryopteris x clandonensis 'Longwood Blue'	Blue Mist spirea	Medium	4	5	Blue	Mid–Late	Full/part	5–9
Ceanothus martinii	Utah mountain-lilac	Low	3	5	White	Early–Mid	Full/part	3–6
Ceanothus velutinus	Snowbrush ceanothus	Medium	6	6	White	Early–Mid	Full/part	3–6
Cercocarpus montanus	Alderleaf mountain mahogany	Low	6	6	Pink	Early–Mid	Full sun	3–6
Chamaebatiaria millefolium	Fernbush	Low	5	5	White	Mid–Late	Full/part	4–9
Chrysothamnus nauseosus 'Tall Blue'	Tall Blue rabbitbrush	Very Low	6	6	Yellow	Late	Full sun	4–9
Cotinus coggygria	Smoketree	Medium	15	10	Pink	Early–Mid	Full sun	4–9
Cotoneaster divaricatus	Spreading cotoneaster	Medium	6	8	White/ pink	Early–Mid	Full/part	4–7
Daphne x burkwoodii 'Carol Mackie'	Carol Mackie Daphne	Medium	3	4	Pink	Early	Full/part	4–8
Eriogonum corymbosum	Lacy buckwheat	Very Low	4	4	Variable	Late	Full sun	3–6
Fallugia paradoxa	Apache plume	Very Low	5	5	White	Mid	Full sun	3–9
Gutierrezia sarothrae	Matchbrush	Very Low	2	2	Yellow	Mid–Late	Full sun	4–9
Hesperaloe parvifolia	Red yucca	Very Low	4	4	Red	Mid–Late	Full sun	5–10
Holodiscus dumosus	Mountain spray	Medium	12	10	White	Mid	Full sun	3–6
Iliamna rivularis	Mountain hollyhock	Medium	6	4	Pink	Mid–Late	Sun or shade	3–6
Jamesia americana	Waxflower	Low-Medium	6	4	White	Early–Mid	Shade	3–6
Lonicera involucrata	Twinberry	Medium	9	10	Yellow/ Pink	Early	Shade	3–6
Peraphyllum ramosissimum	Squaw apple	Low	5	5	White	Early	Full sun	3–6
Philadelphus lewisii 'Cheyenne'	Cheyenne mockorange	Low	9	8	White	Early–Mid	Full/part	3–8

Philadelphus microphyllus	Littleleaf mockorange	Medium	6	4	White	Early–Mid	Full sun	4–10
Physocarpus malvaceus	Mallow-leaved ninebark	Medium	6	4	White/ Pink	Early–Mid	Shade	2–5
Physocarpus opulifolius	Ninebark	Medium	8	6	White	Early–Mid	Full/part	2–8
Potentilla fruticosa	Shrubby cinquefoil	Medium	4	4	Yellow	Mid	Full/part	2–7
Prunus besseyi	Sand cherry	Low	6	6	White	Early	Full/part	3–8
Prunus x cistena	Purple-leaf sand cherry	Medium	10	8	White	Early	Full sun	2–8
Prunus virginiana	Chokecherry	Medium	15	10	White	Early	Full sun	2–6
Rhus glabra var. *cismontana*	Dwarf smooth sumac	Low	5	4	N/A	N/A	Full sun	3–6
Rhus trilobata	Squawbush	Low	4	6	N/A	N/A	Full/part	4–7
Rhus typhina	Staghorn sumac	Medium	15	10	N/A	N/A	Full/part	3–9
Ribes alpinum	Alpine currant	Medium	6	6	Green-yellow	Early	Shade/ part	2–7
Ribes aureum	Golden currant	Medium	6	5	Yellow	Early	Sun or shade	3–6
Rosa glauca	Redleaf rose	Medium	6	5	Pink	Early–Mid	Full sun	2–8
Rosa rugosa	Rugosa rose	Medium	6	6	Variable	Mid–Late	Full sun	3–7
Rosa woodsii	Woods' rose	Low	6	5	Pink	Mid	Full/part	3–6
Shepherdia argentea	Silver buffaloberry	Medium	12	10	N/A	N/A	Full sun	3–9
Sorbus scopulina	Rock mountain-ash	Medium	10	8	White	Mid	Sun/part shade	2–5
Symphoricarpos orbiculatus	Coralberry	Medium	5	8	Pink/ white	Mid	Full/part	2–7
Symphoricarpos oreophilus	Mountain snowberry	Medium	5	5	White/ pink	Mid	Full/part	3–6
Symphoricarpos x chenaultii 'Hancock'	Dwarf snowberry	Medium	2	1	Pink	Mid	Full/part	4–7
Teucrium chamaedrys	Germander	Low	1	2	Pink	Mid–Late	Full sun	4–10
Viburnum trilobum	Cranberry bush	Medium	10	12	White	Early–Mid	Full/part	2–7
Zizophora clinopodioides	Blue mint bush	Low	15	15	Blue	Mid–Late	Full/part	5–9

Ground covers

LATIN NAME	COMMON NAME	WATER-USE RATING	MAX. HT. (FT)	MAX. SPREAD (FT)	FLOWER COLOR	BLOOM PERIOD	SUN/SHADE	HARDINESS ZONE
Achillea tomentosa	Woolly yarrow	Low	0.5	2	Yellow	Mid	Full sun	3–10
Antennaria microphylla.	Rosy pussytoes	Medium	0.25	Spreading	Pink	Mid	Full/part	2–6
Arctostaphylos uva-ursi	Bearberry or kinnikinnick	Medium	1	6	Pink	Mid	Shade/part	3–7
Cerastium tomentosum	Snow-in-summer	Low	0.75	2.5	White	Early	Full/part	3–10
Erigeron flagellaris	Trailing daisy	Low	1	Spreading	White	Mid	Full sun	2–10
Genista lydia	Hardy dwarf broom	Low	1	4	Yellow	Early	Full/part	4–10
Juniperus communis	Common juniper	Very Low	3	1	N/A	N/A	Sun or shade	1–5
Juniperus horizontalis 'Blue Chip'	Creeping juniper	Low	0.75	8	N/A	N/A	Full sun	3–9
Mahonia repens	Creeping Oregon grape	Medium	1	4	Yellow	Early	Shade	4–7
Pachystima myrsinites	Mountain lover	Medium	3	5	Dark pink	Mid	Shade	3–6
Petrophytum caespitosum	Tufted rockmat	Very Low	0.5	3	Cream	Early	Sun or shade	3–7
Phlox hoodii	Carpet phlox	Very Low	0.5	Spreading	Variable	Early	Full sun	3–7
Prunus besseyi 'Pawnee Butte'	Pawnee Butte sand cherry	Low	1.5	6	White	Early–Mid	Full sun	4–8
Rhus aromatica 'Gro-Low'	Gro-Low sumac	Low	3	8	N/A	N/A	Full/part	4–9
Salvia dorrii	Dorr sage	Very Low	2.5	3	Blue/ purple	Early	Full sun	3–6
Sedum kamtschaticum	Russian stonecrop	Low	1	2	Orange-yellow	Mid	Full/part	3–9
Sedum spurium	Dragon's blood stonecrop	Low	0.5	1.5	Pink	Mid	Full/part	3–9
Sedum spurium 'Tricolor'	Variegated stonecrop	Low	0.5	1.5	Pink	Mid	Full/part	2–9
Tanacetum densum ssp. *amani*	Partridge feather	Low	0.5	2	Yellow	Mid	Full/part	4–9
Thymus lanuginosus	Woolly thyme	Low	0.25	3	Pink (rare)	Mid	Full sun	4–9
Thymus praecox 'Coccineus'	Red creeping thyme	Low	0.25	1.5	Red	Early–Mid	Full sun	2–9
Veronica liwanensis	Turkish speedwell	Low	0.25	1.5	Blue	Early–Mid	Shade in afternoon	4–8

Perennials

LATIN NAME	COMMON NAME	WATER-USE RATING	MAX. HT. (FT)	MAX. SPREAD (FT)	FLOWER COLOR	BLOOM PERIOD	SUN/SHADE	HARDINESS ZONE
Achillea filipendulina 'Gold Plate' or 'Coronation Gold'	Fernleaf yarrow	Low	3–5	3	Yellow	Early–Mid	Full sun	3–10
Achillea millefolium 'Moonshine'	Moonshine yarrow	Low	2	2	Yellow	Mid–Late	Full sun	3–9
Aethionema schistosum	Persian stonecress	Low	1	1	Pink	Mid	Full sun	4–8
Agastache rupestris	Licorice mint	Low	3.5	1.5	Variable	Mid	Full/part	5–9
Agastache x 'Desert Sunrise'	Hybrid hummingbird mint	Low	4	2	Variable	Mid–Late	Full sun	5–10
Alchemilla mollis	Lady's mantle	Medium	1.5	2.5	Green	Mid	Full/part	3–8
Allium acuminatum	Tapertip onion	Low	1	0.25	Pink/Purple	Early–Mid	Full sun	3–7
Anacyclus depressus	Mount Atlas daisy	Low	0.25	1.5	White	Early	Full sun	5–9
Anaphalis margaritacea	Pearly everlasting	Medium	2.5	2	White	Mid	Full/part	3–8
Arenaria macradenia	Shrubby sandwort	Low	1.5	2	White	Early–Mid	Full/part	3–8
Armeria maritima	Sea thrift	Low-Medium	1	1	Variable	Early–Mid	Full sun	4–8
Artemisia frigida	Fringed sage	Low	1.5	1.5	Yellow	Mid–Late	Full sun	3–6
Artemisia ludoviciana 'Silver King'	White sage	Low	3	3	Yellow	Mid–Late	Full sun	4–9
Artemisia schmidtiana 'Silver Mound'	Silver mound wormwood	Low	1	1.5	Inconspicuous	N/A	Full sun	3–7
Artemisia versicolor 'Sea Foam'	Sea foam artemisia	Low	0.75	2	Silver-yellow	Occasional	Full/part	4–10
Artemisia x 'Powis Castle'	Silver sage	Low	3	2.5	N/A	N/A	Full/part	4–9
Asclepias tuberosa	Butterfly weed	Low	3	2	Orange	Mid	Full/part	3–9
Aster novae-angliae 'Purple Dome'	Purple dome New England aster	Medium	1.5	1.5	Purple	Mid–Late	Full sun	3–8
Aster x frikartii 'Mönch' or 'Wonder of Staffa'	Mönch hybrid aster	Medium	3	2	Purple	Mid–Late	Full/part	5–9

Astragalus filipes	Basalt milkvetch	Very Low	1.5	1.5	Cream	Early–Mid	Full sun	5–8
Aurinia saxatilis	Basket-of-gold	Low	1	1.5	Yellow	Early–Mid	Full/part	4–7
Baileya multiradiata	Desert marigold	Low	1	2	Yellow	Early–Late	Full/part	Annual in IMW
Bergenia cordifolia	Heartleaf bergenia	Medium	1.5	1.5	Pink/lilac	Early	Shade	3–8
Berlandiera lyrata	Chocolate flower	Low	3	1	Yellow	Early–Late	Full sun	4–8
Callirhoe involucrata	Poppy mallow	Low	0.5	2.5	Magenta	Mid–Late	Full sun	4–9
Calylophus lavandulifolius	Puckered sundrops	Very Low	0.5	1	Yellow	Early–Mid	Full sun	3–10
Calylophus serrulatus 'Prairie Lode'	Prairie Lode sundrops	Low	0.5	1.5	Yellow	Mid–Late	Full/part	3–9
Castilleja integra	Orange Indian paintbrush	Low	1	0.5	Orange	Early–Mid	Full sun	4–7
Catananche caerulea	Cupid's dart	Medium	2	1	Blue/violet	Mid–Late	Full sun	4–7
Centaurea montana 'Blue'	Mountain blue cornflower	Very Low	2	2	Blue/violet	Early–Mid	Full sun	3–8
Centranthus ruber 'Coccineus'	Jupiter's beard	Low	3	2.5	Rose-red	Early–Mid	Full/part	4–9
Coreopsis grandiflora 'Mayfield Giant'	Mayfield giant tickseed	Low	4	1.5	Yellow	Early–Mid	Full sun	4–10
Coreopsis grandiflora 'Sunray' or 'Zagreb'	Tickseed	Low	2	1.5	Yellow	Early–Late	Full sun	3–9
Coreopsis verticillata 'Moonbeam'	Threadleaf tickseed	Low	2	2	Yellow	Mid	Full sun	3–9
Corydalis lutea	Corydalis	Medium	1.5	1.5	Yellow	Early–Late	Shade	5–7
Cytisus purgans	Spanish gold hardy broom	Very Low	4	6	Yellow	Early–Mid	Full/part	4–9
Dalea ornata	Western prairie clover	Very Low	1.5	2	Purple	Mid	Full sun	4–8
Dalea searlsiae	Searls' prairie clover	Very Low	2	2	Purple	Mid	Full sun	4–8
Delosperma ashtonii 'Blut'	Blut iceplant	Very Low	1	1.5	Magenta	Mid	Full/part	5–7
Delosperma congestum 'Gold Nugget'	Gold Nugget iceplant	Low	0.5	0.75	Yellow	Mid–Late	Full/part	5–7

Delosperma cooperi	Hardy purple iceplant	Low	0.25	1.5	Purple	Mid–Late	Full sun	5–7
Dianthus gratianopolitanus 'Fire Witch'	Fire Witch dianthus	Low	0.75	1.5	Pink	Early–Mid	Full/part	3–9
Echium amoenum	Red feathers	Low	1.5	1	Red	Early–Late	Full/part	3–9
Erigeron linearis	Yellow daisy	Very Low	0.5	1.5	Yellow	Early–Mid	Full sun	5–8
Erigeron pulcherrimus var. *pulcherrimus*	Basin fleabane	Very Low	1	0.5	White/Lavender	Early–Mid	Full sun	5–8
Eriogonum racemosum	Pink smoke buckwheat	Very Low	3	0.5	Pink	Late	Full sun	3–9
Eriogonum umbellatum	Sulfur buckwheat	Very Low	1	1.5	Variable	Variable	Full sun	3–9
Eriogonum wrightii var. *wrightii*	Snow Mesa buckwheat	Very Low	1.5	2	White	Mid–Late	Full sun	4–9
Gaillardia aristata	Blanketflower	Low	1.5	1	Yellow/Red	Mid	Full sun	3–10
Gaillardia pinnatifolia	Hopi blanketflower	Low	1.5	1	Yellow/Red	Mid–Late	Full sun	3–10
Gaillardia x grandiflora 'Arizona Sun'	Arizona sun blanketflower	Very Low	1	1	Red/Yellow	Mid–Late	Full sun	3–9
Gaura lindheimeri	Whirling butterflies	Low	5	2	White/Pink	Mid–Late	Full sun	5–9
Gaura lindheimeri 'Siskiyou Pink'	Siskiyou Pink gaura	Medium	1.5	1.5	Pink	Mid–Late	Full sun	5–8
Genista tinctoria	Dyer's broom	Low	3	2	Yellow	Early–Mid	Full sun	4–7
Geranium viscosissimum	Sticky geranium	Medium	2	2	Pink	Early–Mid	Part shade	2–8
Gypsophila paniculata 'Bristol Fairy'	Baby's breath	Medium	3	3	White	Early–Mid	Full sun	3–9
Helenium hoopesii	Orange sneezeweed	Medium	3	2	Yellow	Mid	Full/part	3–8
Helianthemum x nummularium 'Cheviot'	Cheviot rockrose	Low	1	3	Peach	Early–Mid	Full/part	5–8
Heuchera sanguinea 'Splendins'	Splendins coral bells	Medium	1.5	1.5	Red	Mid	Part shade	5–9
Hymenoxys acaulis	Sundancer daisy	Very Low	1	0.5	Yellow	Early–Mid	Full sun	4–10
Iberis sempervirens	Candytuft	Medium	1	1.5	White	Early	Full sun	3–8
Iris pallida 'Variegata'	Variegated sweet iris	Low	2.5	1	Violet-blue	Early–Mid	Full sun	4–9
Kniphofia uvaria 'Hybrid Mix'	Red hot poker	Low	3	3	Orange/yellow	Early–Mid	Full/part	4–10
Lavandula angustifolia 'Hidcote'	English lavender	Low	3	1.5	Purple	Early–Mid	Full/part	5–8

Liatris punctata	Gayfeather	Low	1.5	1	Purple	Mid–Late	Full sun	4–8
Liatris spicata	Button snakewort	Low	4	1.5	Purple	Mid–Late	Full sun	3–8
Limonium latifolium 'Blue Cloud'	Blue Cloud lavender	Low	2	2	Purple	Mid–Late	Full sun	4–9
Linum lewisii	Blue flax	Very Low	2.5	2.5	Blue	Early–Mid	Full sun	4–9
Marrubium rotundifolia	Silver edged horehound	Low	1	1.5	Inconspicuous	N/A	Full sun	5–9
Melampodium leucanthum	Blackfoot daisy	Medium	1.5	2	White	All Season	Full sun	5–9
Mirabilis multiflora	Showy four o' clock	Low	1.5	4	Magenta	Mid–Late	Full/part	4–8
Monardella odoratissima	Mountain beebalm	Low	1	1	Purple	Mid–Late	Full sun	3–10
Nepeta x faassenii 'Blue Wonder'	Blue catmint	Low	1.5	1.5	Purple	Mid–Late	Full/part	3–8
Oenothera caespitosa	White-tufted evening primrose	Very Low	1	2	White	Early–Mid	Full sun	4–9
Oenothera howardii	Bronze evening primrose	Very Low	2	1	Yellow	Early–Mid	Full sun	4–9
Oenothera macrocarpa	Missouri evening primrose	Very Low	1	2	Yellow	Early–Mid	Full sun	4–8
Oenothera pallida	Pale evening primrose	Very Low	2	1	White	Early–Mid	Full sun	4–9
Origanum libanoticum	Hopflower oregano	Low	1.5	1.5	Pink	Mid–Late	Full/part	5–10
Penstemon eatonii	Firecracker penstemon	Low	3	1.5	Red	Early–Mid	Full sun	4–9
Penstemon grandiflorus 'Prairie Jewel'	Prairie Jewel penstemon	Low	3	1	Variable	Early–Mid	Full sun	4–9
Penstemon palmeri	Palmer penstemon	Very Low	5	2	Pink	Early–Mid	Full sun	4–9
Penstemon pinifolius 'MerseaYellow'	Yellow pineleaf penstemon	Low	1	1	Yellow	Early–Mid	Full/part	4–9
Penstemon rostriflorus	Bridges' penstemon	Very Low	2	2	Orange-red	Mid–Late	Full/part	4–9
Penstemon strictus	Rocky Mountain penstemon	Low	2	3	Purple	Early–Mid	Full/part	4–9
Penstemon x mexicali 'Pikes Peak Purple'	Pikes Peak Purple beardtongue	Low	1.5	1.5	Purple	Early–Mid	Full/part	5–9
Penstemon x mexicali 'Red Rocks'	Red Rocks hybrid beardtongue	Low	1.5	1.5	Pink/white	Early–Mid	Full/part	5–9

Perovskia atriplicifolia	Russian sage	Low	4	3	Purple	Mid–Late	Full sun	4–9
Phlomis cashmeriana	Kashmir sage	Low	4	2	Purple	Mid	Full/part	5–9
Phlomis russeliana	Hardy Jerusalem sage	Low	3	2	Yellow	Mid	Full/part	4–10
Phlox grayi	Arizona pink phlox	Low	0.5	1.5	Pink	Early–Mid	Full sun	4–8
Phlox pilosa ssp. *sangamonensis*	Sangamon River prairie phlox	Low	1	2.5	Purple	Early–Mid	Full sun	5–8
Potentilla gracilis	Showy cinquefoil	Low	2	2	Yellow	Mid	Full sun	3–10
Pulsatilla vulgaris	European pasqueflower	Medium	1	1	Purple	Early	Full/part	4–8
Ratibida columnifera 'Mexican Hat'	Prairie coneflower	Low	2	1.5	Red	Mid	Full sun	4–8
Rudbeckia fulgida 'Goldsturm'	Black-eyed Susan	Medium	2	2	Yellow	Mid–Late	Full sun	4–9
Salvia argentea	Silver sage	Low	1.5	2	Grown for foliage	N/A	Full sun	5–9
Salvia greggii 'Furman's Red'	Furman's Red sage	Low	2	1.5	Red	Early–Late	Full/part	5–10
Salvia nemorosa 'East Friesland'	East Friesland meadow sage	Low	1.5	1	Purple	Mid–Late	Full sun	4–8
Salvia nemorosa 'May Night'	May Night meadow sage	Low	1.5	1.5	Purple	Early–Mid	Full sun	4–9
Salvia officinalis 'Minima'	Little-leaf culinary sage	Low	1	1.5	Blue	Mid	Full/part	5–9
Santolina chamaecyparissus 'Virens'	Green santolina	Low	4	2	Yellow	Mid–Late	Full sun	5–10
Scabiosa columbaria 'Butterfly Blue'	Butterfly Blue pincushion flower	Low	1.5	1.5	Lavender-blue	Mid–Late	Full/part	5–9
Sedum middendorffianum	Chinese mountain stonecrop	Low	0.5	1.5	Yellow to orange-red	Mid–Late	Full sun	4–8
Sedum sieboldii	October Daphne stonecrop	Low	0.5	1	Pink	Late	Full/part	3–10
Sedum spectabile 'Autumn Joy'	Autumn Joy sedum	Medium	1.5	1.5	Pink to bronze	Late	Full sun	3–8
Sphaeralcea coccinea	Scarlet globemallow	Very Low	1.5	1.5	Orange-red	Early–Mid	Full sun	4–9
Stanleya pinnata	Prince's plume	Low	4	2	Yellow	Mid	Full sun	4–8

Veronica spicata 'Blue Charm'	Spiked speedwell	Medium	1.5	1.5	Blue	Mid	Full sun	3–8
Viguiera multiflora	Showy goldeneye	Medium	0.5	1	Yellow	Mid–Late	Full sun	3–8
Zauschneria garrettii 'Mountain Flame'	Mountain Flame fire chalice	Low	1.5	2	Red-orange	Mid	Full/part	5–9
Zauschneria latifolia	Fire chalice	Low	1.5	2	Red	Mid–Late	Full sun	5–9
Zinnia grandiflora	Prairie Zinnia	Low	0.5	1.5	Yellow	Mid–Late	Full/part	4–9

Cacti & Succulents

LATIN NAME	COMMON NAME	WATER-USE RATING	MAX. HT. (FT)	MAX. SPREAD (FT)	FLOWER COLOR	BLOOM PERIOD	SUN/SHADE	HARDINESS ZONE
Agave utahensis	Dwarf Utah century plant	Very Low	1	1	Yellow	Mid	Full sun	5–10
Echinocereus engelmannii	Hedgehog cactus	Very Low	2	3	Magenta	Early–Mid	Full sun	5–11
Echinocereus triglochidiatus	Claretcup cactus	Very Low	1	3	Red	Early–Mid	Full sun	5–11
Opuntia basilaris	Beavertail cactus	Very Low	1	4	Pink	Early– Mid	Full sun	5–11
Opuntia erinacea	Mojave pricklypear	Very Low	1	3	Variable	Early–Mid	Full sun	5–10
Opuntia polyacantha	Plains pricklypear	Very Low	1	2	Variable	Early–Mid	Full sun	5–10
Pediocactus simpsonii	Mountain cactus	Very Low	0.5	1	Variable	Early–Mid	Full sun	4–8
Yucca harrimaniae	Harriman yucca	Very Low	3	3	Cream	Early–Mid	Full sun	4–8

Vines

LATIN NAME	COMMON NAME	WATER-USE RATING	MAX. HT. (FT)	MAX. SPREAD (FT)	FLOWER COLOR	BLOOM PERIOD	SUN/SHADE	HARDINESS ZONE
Campsis radicans 'Flamenco'	Scarlet trumpet vine	Medium	40	3	Red/orange	Mid–Late	Full/part	4–10
Campsis radicans 'Flava'	Yellow trumpet vine	Medium	40	3	Yellow	Mid–Late	Full/part	5–9
Campsis x tagliabuana 'Madame Galen'	Trumpet creeper	Medium	25	12	Salmon	Mid	Full/part	4–9
Clematis ligusticifolia	Western white clematis	Medium	20	2	White	Mid	Full/part	3–7
Clematis tangutica	Golden clematis	Medium	15	2	Yellow	Mid–Late	Full sun	4–9
Wisteria floribunda	Japanese wisteria	Medium	30	2	Purple	Early–Mid	Full sun	5–9

Ornamental Grasses

LATIN NAME	COMMON NAME	WATER-USE RATING	MAX. HT. (FT)	MAX. SPREAD (FT)	FLOWER COLOR	BLOOM SEASON	SUN/SHADE	HARDINESS ZONE
Achnatherum hymenoides	Indian ricegrass	Low	2	2	Seedhead	Mid–Late	Full sun	3–8
Achnatherum speciosum	Desert needlegrass	Very Low	2	2	Seedhead	Mid	Full sun	5–10
Andropogon gerardii	Big bluestem	Low	6	4	Seedhead	Mid–Late	Full sun	2–8
Aristida purpurea	Purple threeawn	Very Low	2	2	Seedhead	Mid–Late	Full sun	5–10
Bouteloua curtipendula	Sideoats grama	Low	3	2	Seedhead	Mid–Late	Full sun	2–8
Calamagrostis x acutiflora 'Avalanche'	Feather reed grass	Medium	5	2	Seedhead	Mid–Late	Full/part	5–9
Calamagrostis x acutiflora 'Karl Foerster'	Karl Foerster feather reed grass	Medium	5	2.5	Seedhead	Early–Late	Full sun	5–9
Danthonia intermedia	Timber oatgrass	Low	1.5	1.5	Seedhead	Mid	Full/part	3–8
Deschampsia caespitosa	Tufted hairgrass	Medium	3	2	Seedhead	Mid–Late	Full/part	4–9
Festuca glauca 'Sea Urchin'	Sea Urchin blue fescue	Medium	0.75	1	Seedhead	Mid–Late	Full/part	4–9
Festuca idahoensis	Idaho fescue	Medium	2	2	Seedhead	Mid	Full/part	2–8
Festuca ovina var.*glauca*	Blue fescue	Medium	1	1.5	Seedhead	Mid	Full sun	4–8
Hesperostipa comata	Needle-and-thread grass	Very Low	4	2	Seedhead	Early-Mid	Full sun	3–8
Koeleria macrantha	Junegrass	Low	2	1	Seedhead	Mid	Full/part	3–9
Leymus cinereus	Great Basin wildrye	Low	5	3	Seedhead	Early-Mid	Full sun	2–8
Miscanthus sinensis 'Yaku Jima'	Yaku Jima maiden-hair grass	Medium	4	4	Seedhead	Late	Full/part	5–9
Nassella tenuissima	Mexican feather grass	Low	3	2	Seedhead	Mid–Late	Full/part	7–11
Panicum virgatum	Switchgrass	Medium	5	3	Seedhead	Mid	Full sun	2–8
Pseudoroegneria spicata	Bluebunch wheatgrass	Low	3	2	Seedhead	Mid	Full sun	3–9
Saccharum ravennae	Hardy pampas grass	Medium	5	4	Seedhead	Mid–Late	Full sun	5–10
Schizachyrium scoparium	Little bluestem	Low	3	2	Seedhead	Mid–Late	Full sun	3–9
Sorghastrum nutans	Indiangrass	Medium	6	2	Seedhead	Mid–Late	Full sun	4–9
Sporobolus airoides	Alkali sacaton	Very Low	3.5	2	Seedhead	Mid–Late	Full sun	4–9

Principle 4

Practical Turfgrass Areas

Kelly Kopp and JayDee Gunnell

Of the guiding principles of water-efficient landscaping, the most controversial involves the use of turfgrass in the landscape. At times it seems as though water-efficient landscaping may not allow the use of turfgrasses at all. In fact, water-efficient landscaping recognizes turfgrass as an integral and beneficial component of the landscape.

The reason that turfgrass is mentioned specifically in water-efficient landscaping guidelines is that there is great potential for overirrigation of turfgrasses. Unlike other plants that readily exhibit the stresses of overwatering, turfgrass is able to withstand a great deal of overirrigation without exhibiting signs of stress. In addition, as an herbaceous plant, turfgrass is often one of the first plants in the landscape to exhibit signs of drought stress. These facts, coupled with a "more is always better" attitude toward landscape irrigation, predispose turfgrass areas to overirrigation.

When meeting the needs of a client for recreation space in the landscape, turfgrass is often the best and only appropriate ground cover choice. Turfgrasses are a very practical landscape surface for many urban and suburban landscapes, but sometimes clients request grasses for locations where the purpose is not recreation. In these cases, lower-water-use plant materials may be suggested and considered. Refer to "Principle 3: Appropriate Plant Selection" for possible options.

As a design component, turfgrass invites participation in the landscape while providing unity and simplicity. When using turfgrass in a water-efficient landscape, a few basic guidelines are helpful:

If the only time a turf area receives traffic is when it is mowed, perhaps a low-maintenance plant would work in that location.

- Use turfgrass only in areas where it is functional. These areas may include play areas, areas receiving traffic, and areas needing temperature, noise, or dust mitigation. If the only time a turf area receives traffic is when it is mowed, perhaps a lower-maintenance plant would work in that location.
- Consider choosing turfgrass species with lower water requirements. Depending on where you are in the region, certain varieties of the different turfgrass species will perform better. Refer to your local university extension service for recommended varieties.
- Consider using nonirrigated turfgrass areas. If the turfgrass is not performing a functional role, does it really need to be irrigated? Many turfgrasses can withstand considerable drought stress by entering dormancy (turning yellow/brown). When conditions improve, they will green up again.
- Do not plant turfgrass in narrow, small, or oddly shaped areas that are difficult to irrigate or mow efficiently. In these types of locations, there are many other plants that are more practical choices. Refer to "Principle 3: Appropriate Plant Selection" for other plant options.
- Hydrozoning in a water-efficient landscape certainly applies to turfgrasses as well as other plants. Plan and design irrigation systems so that turfgrass areas are irrigated separately from other landscape plants. Also, become familiar with the actual water requirements of the turfgrass, and set irrigation controllers so that those requirements are not exceeded (Table 4-1).
- Use cultural practices that will improve turfgrass water-use efficiency. For example, mowing higher will encourage deeper rooting and will improve heat- and drought-tolerance. Proper fertilization will also support healthy turfgrass and allow it to better withstand the stresses of heat and drought. Returning grass clippings when mowing helps to reduce evaporation of water from the soil surface.

When these guidelines are followed, turfgrass becomes an appropriate, practical, and beautiful component of the water-efficient landscape.

Turfgrass Irrigation

In the Intermountain West, proper turfgrass irrigation is essential for the health of the grass and will also help to conserve limited water supplies. The following practices should be followed to optimize irrigation efficiency in turfgrass areas:

- Set schedules to irrigate in the early morning when temperatures are cool, which will allow the grass to dry quickly and lessen its disease susceptibility.
- Avoid irrigating between 10:00 a.m. and 6:00 p.m. to minimize evaporation losses.
- Encourage clients to check irrigation systems monthly for leaks and damaged or clogged sprinkler heads.
- *Encourage clients to accept a less-than-perfect lawn*. During drought, grass will take on a golden yellow color as it goes dormant, but will recover as temperatures cool and water becomes more available. Avoid excessive traffic on dormant turfgrass.
- When puddles form or water runs off during irrigation, set irrigation schedules to water in short, repeated cycles. Schedules should allow one hour between cycles for water to infiltrate into the soil.
- Encourage clients to adjust irrigation controllers seasonally to meet the changing water requirements of the grass.

Remember that subsurface drip lines much be buried at a depth (at least six inches) and spacing (twelve to twenty-four inches) that will allow for routine turgrass maintenance practices such as aeration.

Turfgrass irrigation is typically accomplished in one of two ways. In smaller areas that may have slightly odd or curved shapes, spray irrigation heads should be used. Spray heads typically apply between one and two inches of water per hour with a radius-of-throw reaching eighteen feet or less. For larger landscapes with mainly straight lines, such as sports fields, rotary sprinkler heads should be used for irrigation. Rotary sprinklers typically apply between one and one and one-half inches of water per hour with a radius-of-throw ranging from fifteen to sixty feet or more.

In some places in the Intermountain West, flood irrigation is still used for landscapes. If the slope of a turfgrass area is appropriate and the soil texture is suitable, this can be an efficient way to irrigate. With soils that are very sandy, however, water may infiltrate too quickly near the water source, and the entire landscape is never uniformly covered with water. This is also true of areas in which the slope of the ground directs water toward some areas but not evenly toward other areas.

One technique that has been explored in recent years for turfgrass irrigation is subsurface drip irrigation. This type of irrigation is certainly an option; however, the amount of drip line required to adequately cover an irrigated turfgrass area may be cost-prohibitive. Subsurface drip line may be a practical option for smaller turfgrass areas that are sloped or bermed and, therefore, difficult to irrigate efficiently using spray or rotary sprinklers. Also, remember that subsurface drip lines must be buried at a depth (at least six inches) and spacing (twelve to twenty-four inches) that will allow for routine turfgrass maintenance practices such as aeration. Subsurface drip irrigation is not a practical choice in landscapes in which the design or plant materials will be changing and requiring a significant adaptation of irrigation systems.

Like a car, an irrigation system requires routine maintenance. A regular schedule for irrigation system inspections will pinpoint problems before they cause damage to the grass, or use water inefficiently. Refer to "Principle 6: Irrigation Planning" for checklists of irrigation system maintenance practices.

Table 4-1. Irrigation requirements for different turfgrass species commonly used in the Intermountain West.

TURFGRASS SPECIES	WATER REQUIREMENTS*
Bermudagrass	Low
Blue Grama	Low
Buffalograss	Low
Crested Wheatgrass	Very Low
Fineleaf Fescue	Moderate/Low
Kentucky Bluegrass	High**
Perennial Ryegrass	High
Tall Fescue	Moderate
Zoysiagrass	Low

* Water requirements (inches per growing season): High (25-30 inches), Moderate (20-25 inches), Low (16-20 inches), Very low (no irrigation required beyond the establishment period).

**The rating for Kentucky Bluegrass assumes that the grass will be watered through the heat of the summer rather than letting it go dormant.

Fertilization Requirements

Soil testing is always a good practice when working with turfgrasses and other landscape plants. A sample may be analyzed for basic soil properties as well as

Fertilization may be reduced by as much as one half of clippings are consistently recycled.

the levels of phosphorus and potassium that are in the soil. If additional phosphorus and potassium are required, the test will provide recommended application rates. Refer to "Principle 2: Soil Preparation" for more detailed instructions regarding soil sampling and analysis options.

Although not analyzed in typical soil tests, nitrogen (N) in most soils of the Intermountain West is insufficient to meet the requirements of turfgrasses. This is particularly true when those grasses are expected to withstand a good deal of traffic. In developing nitrogen fertilization schedules, consider the amount and type of use that the area will receive. Sports fields and parks that receive heavy traffic will benefit from higher nitrogen levels that will aid in their recovery. Home lawns and lesser-used areas can be maintained at acceptable levels of quality with more moderate nitrogen rates. For very low-use areas, a small amount of nitrogen will help maintain the density of the grass over time and prevent weed encroachment (Tables 4-2 and 4-3).

Table 4-2. Fertilization (N) requirements for cool-season grasses by maintenance level.

MAINTENANCE LEVEL	SPRING APRIL/MAY*	SUMMER JUNE	SUMMER JULY	FALL SEPT	FALL NOV	TOTAL
	------------pounds of N per 1,000 square feet------------					
Low	1	-	-	1	-	2
Medium	1	-	-	1	1	3
High	1	0.5	0.5	1	1	4

Table 4-3. Fertilization (N) requirements for warm-season grasses by maintenance level.

MAINTENANCE LEVEL	EARLY SUMMER EARLY JUNE*	SUMMER JULY	TOTAL
	-----------pounds of N per 1,000 square feet------------		
Low	1	-	1
Medium	1	1	2
High	1	1	2

Also keep in mind that recycling grass clippings into the turfgrass area whenever possible and practical is the best way to keep nutrients in the system. Continually removing clippings removes nitrogen from the turfgrass/soil system. In fact, fertilization may be reduced by as much as one half if clippings are consistently recycled.

For most fertilizations during the year, a slow-release form of nitrogen is preferable to a quick-release form. The exception to this recommendation is the late fall nitrogen fertilization for cool-season grasses. This should be applied after the last mowing of the season, while the grass is still green but is no longer actively growing. At that time, a quick-release form of nitrogen should be applied. This will allow the grass to take up the nitrogen and store it over the winter. Green-up will be quicker in the spring and drought tolerance will be improved over the following summer.

For reducing the need for turfgrass irrigation, there are few practices that work as well as mowing the grass at the upper end of the recommended height range.

Remember that adequate nitrogen fertilization will help sustain the quality of turfgrass; however, excessive nitrogen fertilization will actually lessen the water-use efficiency of the grass. Try to strike a balance between acceptable quality and the most efficient use of water.

Mowing

Consider mowing to be one routine turfgrass management practice that performs many functions. Regular mowing helps to improve the density and texture of turfgrasses. Mowing at the upper end of the recommended height range for a given turfgrass species provides the additional benefits of shading the soil and preventing weed seeds from germinating (Table 4-4).

For reducing the need for turfgrass irrigation, there are few practices that work as well as mowing the grass at the upper end of the recommended height range. The relationship between mowing height and rooting depth is well understood. The higher a grass is mowed, the deeper the roots will grow into the soil. The deeper the roots grow into the soil, the more soil volume they will be able to access in order to take up water, and the less frequently irrigation will be required. Deeper roots also strengthen the turfgrass, allowing it to withstand stresses of other kinds such as disease or insect pressure.

As previously mentioned, when mowing, recycle grass clippings back onto the lawn whenever possible. Clippings contain up to 5% nitrogen and they will quickly decompose to provide an additional source of fertilization for the grass.

Table 4-4. Mowing heights by species.

TURFGRASS SPECIES	MOWING HEIGHT RANGE (INCHES)*
Bermudagrass	0.125–1.0
Blue Grama	0.5–3.0
Buffalograss	0.5–unmowed
Crested Wheatgrass	2.0–3.0
Fineleaf Fescue	1.5–3.0
Kentucky Bluegrass	0.75–3.0
Perennial Ryegrass	1.0–3.0
Tall Fescue	2.5–3.0
Zoysiagrass	0.5–1.5

*Mowing at or near the higher end of the mowing height range will encourage deeper rooting and improve turfgrass water-use efficiency.

Aeration

Over time, the ability of some turfgrass areas to absorb applied irrigation water may decrease. This may be the result of soil compaction due to traffic, or may also occur naturally over time in finer-textured soils. In either case, aeration (also known as cultivation or aerification) is a practice that can improve the movement of water into the soil beneath compacted turfgrass areas.

Another condition that may be remedied with aeration is the buildup of excessive thatch levels. Thatch, which comprises both living and dead roots and stems of turfgrass, naturally occurs between the growing grass tissue and the soil beneath it. Some amount of thatch is perfectly normal, and even serves to cushion and protect the grass. However, as thatch depths increase beyond one inch, there is the potential for the grass to begin rooting directly into the thatch layer. If that happens, the grass remains in a constant state of drought stress because the thatch layer is very light and incapable of retaining adequate moisture.

Turfgrass aeration can help to remedy both compacted soils and excessive thatch layers. Hollow-tine, or core aeration, is preferable to solid-tine aeration because it causes less compaction. Although aeration tines vary in length, the goal should be to achieve penetration into the soil of at least four inches. The cores of soil and turfgrass tissue that remain on the ground surface may be picked up if necessary for the functionality of the turfgrass, but it is preferable to allow them to disintegrate and filter down to the soil surface. For more detailed information on turfgrass aeration, see "Turfgrass Cultivation (Aerification)," Utah State University Extension Publication HG/Turf/2006-02, extension.usu.edu/publications.

Turfgrass aeration can help to remedy both compacted soils and excessive thatch layers.

Another practice that can be used in cases of extreme thatch buildup is de-thatching. De-thatching is commonly used on golf courses and some other intensively managed sports fields; however, *it is almost never warranted for a residential landscape*. The practice of de-thatching does indeed pull large amounts of thatch from beneath the turfgrass, but it also pulls out healthy roots and may leave the grass extremely stressed and vulnerable to many other environmental pressures. In addition, the large amount of material that is pulled out must be removed to prevent smothering the turfgrass beneath it.

Turfgrass aeration does cause some stress to the turfgrass and should, therefore, be performed at times of the year when the grass is extremely healthy and actively growing. For cool-season grasses (Kentucky bluegrass, tall fescue), this will be during either the fall or the spring, or perhaps both. For warm-season grasses (buffalograss), this will be during the heat of the summer.

Design Concerns

Landscape design issues are critical when creating and managing turfgrass areas. Small, oddly shaped areas are difficult to irrigate and are often better suited to plant materials other than turfgrass. Also keep in mind that trees planted in turfgrass areas will suffer due to competition for nutrients and water. Trees may also suffer from injuries incurred when mowers and trimmers for turfgrass are used in close proximity to tree trunks (see "Principle 5: Use of Mulch").

Species Information (see Table 4-5)

Cool-season grasses are tolerant of cold temperatures. They tend to green up in early spring and are slow to go dormant in the fall. They need frequent watering in the heat of summer, though, to maintain their health and appearance. Warm-season grasses are slower to green up in spring, but they are very drought-tolerant and look their best in the heat of summer. They tend to go dormant sooner in the fall than cool-season grasses, as temperatures drop.

All grass drawings by Barbara Middleton

Cool-season. The leaf blades of Kentucky bluegrass are fine-textured with a blue-green color. The grass is recognizable by its boat-shaped leaf tip and visible midrib on the top of the blade.

Kentucky bluegrass *(Poa pratensis L.*, native to Eurasia)

Kentucky bluegrass is used far more than any other grass species in the Intermountain West, and for good reason. It is a fine-bladed, rhizomatous grass with a deep, blue-green color and a soft texture. In addition, it is extremely traffic-tolerant and recovers well from injury and stress including drought. Seeding rates for Kentucky bluegrass are 1 to 1.5 pounds per 1,000 square feet.

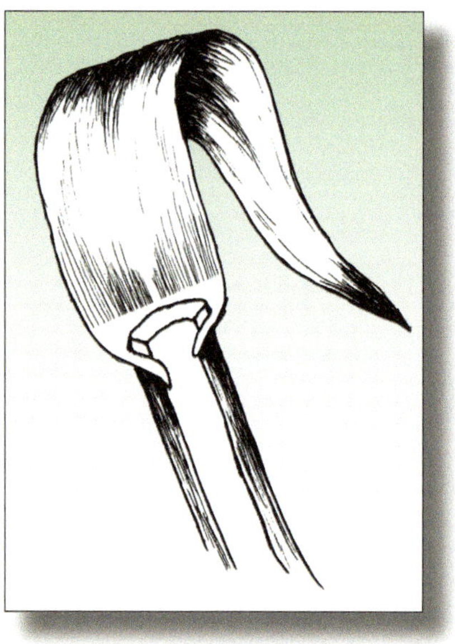

Cool-season. The tall fescues have rough-textured blades and edges. The papery membrane (ligule) at the base of the leaf blade is short and very blunt.

Tall fescue (*Festuca arundinacea Schreb.*, native to Europe)

Turf-type tall fescues have resulted from years of breeding programs working with forage-type tall fescues. These days, the available varieties of turf-type tall fescues have great color and a much finer texture than the forage types. They differ from bluegrass in that they have a bunch-type growth habit and are more deeply rooted, but are otherwise quite similar. Appropriate seeding rates are critical for achieving a dense stand of tall fescue or any other bunch-type grass. Seeding rates for tall fescue are 7 to 9 pounds per 1,000 square feet.

Perennial ryegrass (*Lolium perenne L.,* native to Asia and North Africa)

Perennial ryegrass is often included in seed mixtures planted in the Intermountain West because of its quick germination rate. Seedlings may be seen in as little as one week after planting. Like tall fescue, perennial ryegrass has a bunch-type growth habit, and seeding rates are critical. Perennial ryegrass is the least cold-tolerant of the cool-season grasses and is generally outcompeted over time by species such as Kentucky bluegrass and tall fescue. Because of its quick germination rate, however, perennial ryegrass is often used on sports fields where quick recovery of injured areas is a must. Seeding rates for perennial ryegrass are 7 to 9 pounds per 1,000 square feet.

Warm-season. Buffalograss leaf blades are light gray-green; they are hairy on both sides and on the edges.

Buffalograss [*Bouteloua dactyloides (Nutt.) Engelm.,* native to North America]

Buffalograss is a stoloniferous, warm-season grass. It is well adapted, however, to many parts of the Intermountain West and is able to withstand cold temperatures, despite its classification as a warm-season grass. This ability allows it to survive in cold areas, but means that green-up is slower in the spring, and dormancy is quicker in the fall than for cool-season grasses. Buffalograss has a gray-green color that is lighter than Kentucky bluegrass or tall fescue, and it may either be mowed to recommended heights or left unmowed, in which case it will reach approximately six inches in height. Unmowed, buffalograss produces very attractive seedheads that move gracefully when the wind blows. Seeding rates for buffalograss are 3 to 6 pounds per 1,000 square feet, although it may also be planted with plugs or sprigs, depending on the variety chosen.

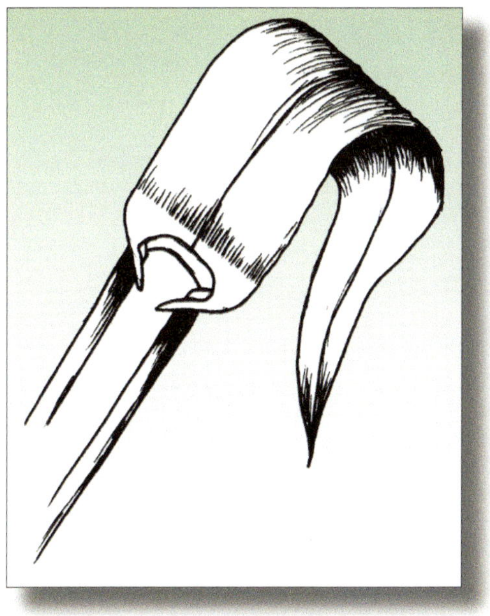

Cool-season. Perennial ryegrass leaf blades are dull above, with distinct ribbing and glossy-shiny below. The leaf tip is pointed. The grass can be recognized by the small, claw-like appendages (auricles) that wrap around the stem at the base of the blade.

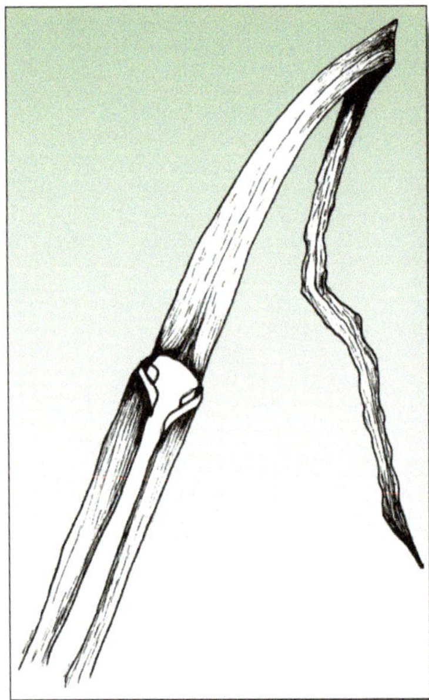

Cool-season. Fineleaf fescue leaf blades are dark green and much narrower than the leaf blades of other turfgrasses.

Blue grama [*Bouteloua gracilis* (Willd. ex Kunth) *Lag. ex Griffiths,* native to North America]

Blue grama is a rhizomatous, warm-season grass. It is well adapted to the Intermountain West and survives cooler temperatures while also having excellent heat- and drought-tolerance. It is often planted in combination with buffalograss, but is best suited to areas that are less intensively managed. Like buffalograss, blue grama has slower green-up in the spring and quicker dormancy in the fall than cool-season grasses. Seeding rates for blue grama are 1.5 to 2.5 pounds per 1,000 square feet.

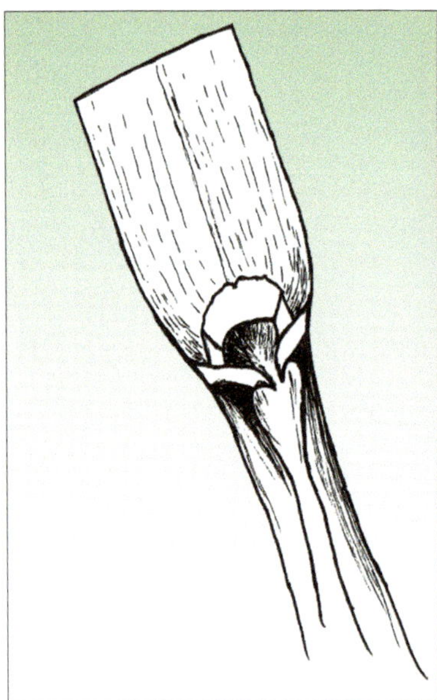

Cool-season. Crested wheatgrass has a long, papery thin membrane (ligule) at the base of the leaf blade. The claw-like appendages (auricles) surrounding the stems are long, similar to those of perennial ryegrass.

Crested wheatgrass [*Agropyron cristatum (L.) Gaertn.,* native to North America]

Crested wheatgrass is a bunch-type, cool-season grass that has a bright green color. Once established, crested wheatgrass will survive the growing season of the cooler Intermountain West areas without supplemental irrigation. This makes it a good choice for areas that need grass cover, but are not irrigated regularly (low intensity management). Seeding rates for crested wheat grass are 3 to 5 pounds per 1,000 square feet.

Fineleaf fescues (F*estuca spp.,* native to Europe)

The fineleaf fescues (including creeping red, Chewing's, sheep, and hard fescues) are bunch-type, cool-season grasses that bear little resemblance to tall fescue. These grasses are very finely bladed (like broom bristles) and exhibit a wide variety of colors depending on the species. The fineleaf fescues do extremely well in shaded areas, and are often included in seed mixtures for sun/shade areas. Seeding rates of fineleaf fescues are 3.5 to 4.5 pounds per 1,000 square feet.

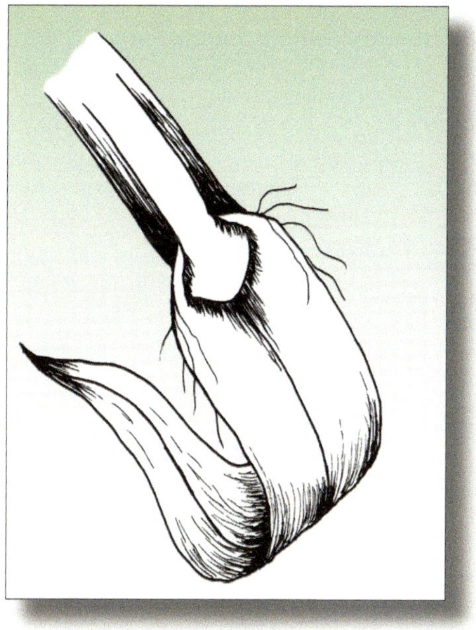

Warm-season. The stems of blue grama grass are round in cross-section. The leaf blades have a broad collar with hairy margins.

Bermudagrass [*Cynodon dactylon (L.) Pers.,* native to Africa]

Bermudagrass is both a stoloniferous and rhizomatous warm-season grass that is well adapted to the warmer areas of the Intermountain West. Its color ranges from a light green to a very dark green depending on the variety. Bermudagrass grows and establishes very quickly and is able, therefore, to recover well from injury. However, this also means that it can be an aggressive weed in some situations. Propagation of bermudagrass is normally by sprigs, plugs, or sod.

Warm-season. The leaf blades of bermudagrass are flat and sharply pointed at the tip. The base of the leaf is hairy, and the stems are round in cross-section.

Warm-season. The stems of zoysiagrass are round in cross-section, and the leaf blades are wide with hairy collars. A fringe of hairs replaces the membranous ligule at the base of the leaf blade.

Zoysiagrasses (*Zoysia spp.*, native to Asia)

Zoysiagrass is both a stoloniferous and rhizomatous warm-season grass that is well adapted to the warmer areas of the Intermountain West and has also survived in some of the cooler areas of the region. Its color is a very attractive dark green. Of potential concern is the very fibrous nature of its leaf blades, which necessitates the consistent use of very sharp mower blades to prevent shredding of its leaf tips and poor mowing quality. All varieties of zoysiagrass are propagated by sprigs, plugs, or sod.

Table 4-5. Characteristics of commonly used turfgrass species in the Intermountain West.

SPECIES	FERTILITY REQUIREMENTS*	DROUGHT TOLERANCE	SHADE TOLERANCE	MOWING QUALITY	WEAR TOLERANCE
Bermudagrass	High	Excellent	Poor	High	Excellent
Blue Grama	Moderate/Low	High	Low	Moderate	Moderate
Buffalograss	Moderate	High	Low	Moderate	Moderate
Crested Wheatgrass	Low	Very High	Low	Poor	Low
Fineleaf Fescue	Low	High	High	Moderate	Moderate
Kentucky Bluegrass	Moderate/High	Moderate	Low	Good	High
Perennial Ryegrass	High	Low	Low	Moderate	High
Tall Fescue	Low	Moderate	Moderate	Good	High
Zoysiagrass	High	Excellent	Good	Poor	High

*Fertility requirements (lb. N per 1,000 ft2 per growing season): High (3–4 lb. N), Moderate (2–3 lb. N), Low (1–2 lb. N).

Principle 5

Use of Mulch

Heidi A. Kratsch and Margaret Shao

Why Mulch?

Mulching around plants in a water-efficient landscape helps keep valuable water in the soil around plant roots and discourages the germination and growth of weed seeds. This not only minimizes irrigation requirements but results in a landscape with low or less frequent maintenance needs and reduces pesticide use. Hoeing or spading around plants to remove weeds can damage plant roots, disrupt soil structure, and discourage the presence of beneficial soil organisms. Mulching won't completely eliminate the need for weed control, but it can decrease the amount of time and labor used for this task. Mulching also moderates soil temperatures and prevents soil compaction. The presence of mulch promotes root development near the soil surface, increasing water, nutrient, and oxygen availability to the plant. Mulching around landscape trees can reduce competition between tree and turfgrass roots for soil water and nutrients. Finally, mulch can be an attractive design component in a water-efficient landscape.

Types of Mulch

Organic mulches include materials such as wood or bark chips, shredded bark, nut shells, pine needles, sphagnum peat, and grass clippings or other discarded disease- and weed-free plant parts. Certain materials may be more available than others on a regional basis, and their availability and price may dictate which

Photo: Heidi Kratsch

Figure 5-1. Wood chips should be composted and free of weed seeds and pesticide residue.

types are used. Organic mulches have the potential to enhance soil structure, increase soil fertility, and increase soil organic matter. Using organic mulch is a great way to recycle yard waste such as pruned woody plant materials and fallen leaves and needles. To ensure adequate aeration and water infiltration and to slow decomposition, make sure mulch particles are larger than the soil particles below (usually larger than one-half inch in diameter). However, some organic materials such as leaves or lawn clippings can mat together and restrict air and water movement into the soil when left whole; they should be chipped or mulched into one- to three-inch pieces before applying as mulch around plant roots.

It is important to be sure recycled plant materials are free from weed seeds, disease-causing organisms, and pesticide and herbicide residues. One way to prevent harm to plants by mulch is to use disease-free plant parts that have not been treated with pesticides. Another way is to compost mulch before use. Composting results in controlled decomposition of organic materials through the activity of microorganisms (bacteria and fungi). When done properly, composting generates enough heat to kill most weed seeds and disease organisms, and will break down some pesticides.

In most cases, use of organic mulch increases the nutrient fertility of the soil below. This decreases the need for application of fertilizer to mulched plants. However, the use of fresh wood or bark as mulch may lead to nitrogen deficiency because microorganisms that decompose these materials use up the nitrogen that would have been available to the plant. This nitrogen loss can be avoided by using composted mulch, or by adding nitrogen at a rate of one-half to one pound per 1,000 square feet prior to applying mulch material.

Over time, organic mulches break down and will need to be replenished. This can be accomplished by adding more mulch over the top of the decomposed mulch material. Always "rough up" the old decomposing mulch before applying new mulch materials to prevent a water-impervious layer from forming between old and new mulch layers. Although the decomposition of old mulch adds to the soil organic matter, it can lead to nitrogen deficiency. Watch plants for telltale signs of this, such as leaf yellowing that starts at the bottom of the plant and is first apparent in the older leaves on the stem. Replace lost nitrogen when necessary using the previously described application rate.

Inorganic mulches consist of nonplant materials and may include a variety of different kinds of rock or gravel. Different sizes and colors of inorganic mulches are available. The decision about which to use depends upon the design of the landscape, the types of plants used, and availability. Examples include gravel or crushed stone, cobblestones, river rock, and lava rock. The texture of inorganic mulch should complement the scale and type of landscape. For example, cobblestones work great when used judiciously among trees and hedges in an expansive formal or naturalized setting, or as a design element in a rock garden, but they could be overwhelming when overused in small spaces. For a natural look, the choice of inorganic mulch should complement the form and color of the rock in the surrounding native environment.

Inorganic mulches provide some of the same benefits as organic mulches, and offer better drainage than most organic mulches for plants that require it (many drought-tolerant native plants). In addition, inorganic mulches do not decompose or need to be replaced frequently. One disadvantage of inorganic mulches is that the soil beneath them is warmer than it is beneath organic mulches. This can be a problem in hot, sunny climates where plant root systems could be damaged by the heat. In climates where summer temperatures regularly exceed 100°F, organic mulches do a better job of retaining soil moisture and protecting the roots of more heat-sensitive plants from temperature extremes. Inorganic mulches also do not improve soil structure or add to soil fertility or organic matter. The latter can be advantageous in very low-water-use landscapes in which the plants used may be weakened by excess nutrients or organic matter in the soil.

Photo: Heidi Kratsch

Figure 5-2. Cobblestones work well in larger landscapes but appear overwhelming in narrow parking strips.

Photo: Heidi Kratsch

Figure 5-3. Small stones or pebbles complement the natural look of a native perennial garden and provide good drainage for drought-resistant plants.

Application of Mulch

Prior to applying mulch, remove debris and weeds from the area, amend the soil as needed (see "Principle 2: Soil Preparation"), and water the soil thoroughly. As mentioned previously, mulch materials should be larger than the particle size of soil amendments. Apply organic mulch to a depth of three to four inches and inorganic mulch to a depth of two to three inches. A one-inch-thick layer of mulch requires about three cubic yards of material per 1,000 square feet of area. Avoid packing the mulch material, and leave at least two inches of mulch-free area around the base of woody plants and moisture-sensitive perennials (like penstemon and some other native plants) to prevent root-collar diseases and rodent damage. Ideally, the mulched area should extend out to or beyond the drip line (edge of the canopy) of woody plants. The best time to apply or replenish mulch is immediately after planting, in the fall or in the spring after the soil has warmed.

Calculating how much mulch is needed:

Area to be mulched (square feet) × Depth of mulch (inches) / 325 = Amount of mulch to apply (cubic yards)

Weed Barriers

A weed barrier may be installed on top of prepared soil before the application of either organic or inorganic mulch. Weed barriers used for landscapes are made of woven polypropylene material (landscape fabric). Water, air, and nutrients will penetrate this woven barrier. We do not recommend the use of landscape fabrics in naturalistic landscapes in which native plants are used. Many native plants have a natural tendency to spread and reseed, which adds to their charm and style in naturalistic landscapes. Landscape fabrics interfere with these tendencies; the desired look of plants drifting and spreading from one area to another is lost when weed barriers are used.

Solid plastic sheets should never be used as a weed barrier in landscape settings as they may generate excessively high root-zone temperatures in direct sun and interfere with water and air exchange with the soil. Plastic also breaks down faster than woven materials, and becomes unsightly if not replaced on a very regular basis. Plastic sheets can girdle trees and promote fungal root diseases.

During installation of a weed barrier, the soil level should be taken down to about the depth of mulch to be applied; this will keep the mulch in place and minimize the unappealing tattering of the weed barrier over time. Apply the weed barrier in strips over the soil, overlapping the strips about four inches. Make a cross-shaped cut in the material where plants will be installed and fold the material under. Once plants have been installed, place mulch over the weed barrier and replace as needed, usually each spring. When landscape fabric is used, plants may develop fine roots within the mulch and above the weed barrier because of the presence of oxygen; this should not be cause for concern and will not inhibit development of deeper, more extensive roots. Decomposed organic mulch will favor the growth of weeds on top of a weed barrier and should be removed before applying new mulch. Under normal conditions, fabric weed barriers may last up to five years.

Photo: Heidi Kratsch

Figure 5-4. Landscape fabric should be applied in strips and tacked down with stakes, if desired, prior to laying down mulch.

Selection of Mulch

Selection of mulch may be based on cost, availability, choice of plants, climate, size of the area, and landscape design. For example, recycled tumbled glass, available for use in some areas, looks stunning in a formal or urban environment where it complements the surrounding architecture, but it can be cost-prohibitive in a large area. Straw or leaves may be readily available but will blow around and make a mess in areas subject to gusty winds. Plants adapted to cooler, high-elevation habitats may thrive in the valleys with moisture-retaining organic mulches, whereas plants adapted to dry, desert habitats may need the drainage provided by inorganic materials.

Table 5-1. Comparison of commonly available mulch materials.

MATERIAL	WEED CONTROL	DRAINAGE QUALITIES	ORNAMENTAL VALUE	OTHER COMMENTS
Organic:				
Shredded wood or wood chips	fair	fair	yes	Compost to avoid nutrient depletion
Sphagnum peat moss	fair	fair	no	Increases acidity of soil; may be difficult to rewet when dry
Grass clippings	poor	poor	no	Use dried and herbicide-free; use a mulching mower
Straw	poor	fair	no	Compost to kill weed seeds; not for windy areas
Shredded bark or bark chips	fair	fair	yes	Compost to avoid nutrient depletion
Leaves	poor	poor, unless chopped before use	yes (especially with fall color)	May blow around in the wind
Pine needles or pine straw	fair	fair	yes; different colors available	Increases acidity of soil; may increase fire hazard near wood fences or walls
Coir fiber	good	good	yes; different colors available	Some reports of plant growth inhibition
Cocoa hulls	good	poor	no	Do not use in shady areas; stays too wet
Inorganic:				
Pea gravel or crushed stone	fair	good	yes	Avoid use around acid-loving plants
Cobblestones	fair	good	yes	Best in a rock garden
Lava rock	fair	good	yes	Black and red are common
Recycled tire rubber (crumb rubber)	fair	fair	yes	Potential for leaching phytotoxic levels of zinc; unpleasant odor at high temperatures; extreme fire hazard.

Principle 6

Irrigation Planning

Graham Hunter, Molly Waters, and Heidi A. Kratsch

Irrigation Systems

Irrigation of a water-efficient landscape starts with designing a system that will most effectively deliver water to plants with the least amount of waste. The following are some key concepts to keep in mind during the design process:

- The design of the landscape, including types and locations of plants, should drive decisions about the style and layout of the irrigation system.
- The irrigation system should be designed to incorporate *hydrozones*. A hydrozone is a distinct area of the landscape defined by the water, drainage, and exposure needs of the plants within. Landscape plants should be grouped (hydrozoned) according to similar needs. Each hydrozone should be irrigated using a separate valve wired either to separate automatic controllers, or to a single multistation controller.
- Parts of the system may be temporary, constructed only for the purpose of establishing very low-water-use plants during their first season or two.
- The system should be timed to irrigate during periods of low evaporation. The best time to irrigate is early in the day.

Two types of landscape irrigation systems exist, and they are uniquely designed to meet the special water needs of different plant types (Table 6-1). Low-volume (drip or microsprinkler) irrigation systems are designed to work at low output pressures, 10 to 30 pounds per square inch (psi) and deliver water directly to the root zones of drought-adapted plants, either individually or in

A hydrozone is a distinct area of the landscape defined by the water, drainage, and exposure needs of the plants within.

groups. Automatic sprinkler systems work at higher pressures (above 40 psi) and volumes, and are effective for irrigation of lawns, some ground covers, and annual flower beds (high-water-use plants). A typical water-efficient landscape may include a combination of sprinkler and low-volume irrigation systems, and each system needs to be zoned separately.

Table 6-1. GALLONS OF WATER NEEDED TO WET THE ROOT ZONE*

Plant water use	Plant canopy diameter (feet) 1	2	3	4	5	6	8	10	12	14
Tree										
medium						21	37	58	83	113
low						15	30	46	66	90
Shrub										
medium		2.5	5	9	14	21	31			
low		2	4	7	12	15	25			
Perennial										
medium	1	3	7							
low	0.5	2	3.5							

* Water requirements were estimated by multiplying the average crown area of a single plant, or the area of a homogeneous planting, by the average weekly reference evapotranspiration (ETo) for north-central Utah during the growing season, and by a plant factor. The plant factor is a dimensionless number expressed as a fraction of ETo and derived based on studies (below) showing that many conventional landscape species maintain their health and appearance when irrigated within a range of 20% to 80% of ETo. For hotter, drier sites in the southern part of the region, multiply these numbers by a factor of 1.4. Very low-water-use plants will need very little to no supplemental irrigation once established.

Pittenger, D. R., D. A. Shaw, D.R. Hodel, and D. B. Holt. 2001. Responses of landscape groundcovers to minimum irrigation. J. Environ. Hort. 19(2):78–84.

Shaw, D. A. and D. R. Pittenger. 2004. Performance of landscape ornamentals given irrigation treatments based on reference evapotranspiration. Acta Hort. 664:607–613.

Staats, D. and J. E. Klett. 1995. Water conservation potential and quality of non-turf groundcovers versus Kentucky bluegrass under increasing levels of drought stress. J. Environ. Hort. 13:181–185.

Most irrigation systems, whether sprinkler or low-volume, share the same basic components that make up the "backbone" of the system, including underground water supply pipes, valves, and controllers. The main differences between conventional sprinkler systems and low-volume systems are the type of emitter used and the pressures (and therefore the volume) of the water flowing through the system. The means of operation of the two systems is essentially the same, and conventional sprinkler systems can be retrofitted to be low-volume, with some minor adjustments to regulate the pressure and prevent clogging of the system. Low-volume systems are no more complicated than conventional sprinkler systems, and have the potential to save water by reducing evaporation and providing water directly to plant roots.

A backflow preventer is required for each zone if culinary (potable) water is used in a private residence.

Basic Irrigation System Components

- Control valves – mechanical devices that either allow or prevent the flow of water to each zone or station in the system.
- Automatic controller – programmable device that controls the timing of water flow to the system. A multistation controller is recommended for ease of programming multiple planting zones.
- Pressure gauge – determines the total pressure available on the system.
- Backflow preventer – prevents backflow from the irrigation system to the water source. A backflow preventer is required for each zone if culinary (potable) water is used in a private residence.
- PVC pipes – transport water from the source to planting areas.
- Emitters – apply water to plants (can be spray heads or drip emitters).

Low-Volume Irrigation

Low-volume irrigation systems are different from conventional higher-volume sprinkler systems because they directly target water application to the plant's root zone. They are low pressure (10–30 psi), and they supply a small volume of water over an extended period of time. Low-volume systems consist of flexible polyethylene tubing (usually one-half inch in diameter) connected to a main line, which is constructed of polyvinyl chloride (PVC) and transports water from the main valve. The flexible tubing can be installed belowground, or aboveground and covered with mulch. An advantage of a low-volume system is that it applies water to a target area and not to unplanted areas, thus reducing weed growth in the space between plants. Low-volume systems work well for oddly shaped and small areas and are perfect for irrigating plants with low water needs. Low-volume systems conserve water by reducing the problems of wind drift and overspray, and by reducing weeds that use up water otherwise available to desirable plants.

Photo: Graham Hunter

Figure 6-1. Point-source drip emitters and fittings.

Added Components for a Low-Volume Irrigation System

- Pressure-compensating – to reduce the water pressure on the main line to between 3 and 30 psi.
- Filter – to remove particles that can clog emitters (some emitters also come with individual filters).
- Polyethylene tubing – half-inch and quarter-inch, to deliver water from PVC pipe main lines to emitters.
- Fittings and end caps – to connect components and cap the ends to maintain water pressure.
- Emitters – Low-volume systems use either drip or pop-up microspray emitters. Pressure-compensating emitters are recommended.

Low-volume irrigation systems employ one of two types of emitters: drip emitters (sometimes called "drippers") and microsprinkler emitters (sometimes called "microsprays"). Drip emitters wet only a small surface area and cover the desired root zone through capillary action. Drip emitters can be either *point-source* or *inline*. Microsprinkler emitters are similar to full-size sprinkler heads, but they operate at lower pressures, cover less area, and are less susceptible than high-pressure sprinklers to wind drift because they do not throw water as high into the air.

Drip System Emitters

Point-source drip emitters (Figure 6-1) are external to the drip line and are punched into the half-inch flexible tubing, either directly by way of a barbed tip on the emitter, or by using a specialized tool. Quality products are pressure-compensating and deliver water at rates between one-half and two gallons per hour. Point-source drip emitters work well for landscapes in which plants (usually small shrubs and herbaceous perennials) are arranged in an irregular pattern with large amounts of space between them (Figure 6-2).

Inline emitters are preinstalled at specific intervals within the half-inch flexible tubing (Figure 6-3). Different spacing is available: emitter spacing intervals of 6, 12, and 20 inches are common. Inline emitters work best with plantings that utilize a grid-pattern design or in a closely spaced flower bed (Figure 6-4). The correct emitter spacing is based on the spacing between plants and the texture of the soil. For example, the diameter of the surface-wetting area will be greater in fine clay soils than in coarse sandy soils, and emitters should be spaced farther apart in clay-type soils (see “Principle 3: Soil Preparation” for more information on soil texture).

Regardless of the type of drip emitter used, it is possible to attach quarter-inch-diameter tubing (also called “spaghetti tubing”) to these emitters for more control over placement within the plant canopy. In this case, a second emitter is attached to the end of the spaghetti tubing, and this emitter remains unburied for ease of inspection and access. Multi-outlet emission devices are available. They are attached to the half-inch tubing and contain several emitters, allowing a central access point to which multiple quarter-inch spaghetti tubing lines can be attached. In this way, larger plant canopies can be efficiently supplied with water by more than one emitter.

Microsprinkler System Emitters

Microsprinkler irrigation is a type of low-volume system that employs small pop-up microspray emitters mounted on stakes to deliver volumes of water that are higher (between 3 and 30 gallons per hour) than that delivered by drip emitters (Figure 6-5). The stakes are connected to half-inch polyethylene tubing by way of quarter-inch spaghetti tubing. The microspray heads have either a fixed or adjustable watering radius, and the spray pattern is either full circle or half circle. Microsprinkler emitters work

Photo: Roger Kjelgren

Figure 6-2. Drip irrigation line with point-source emitter targeting water to the plant root zone.

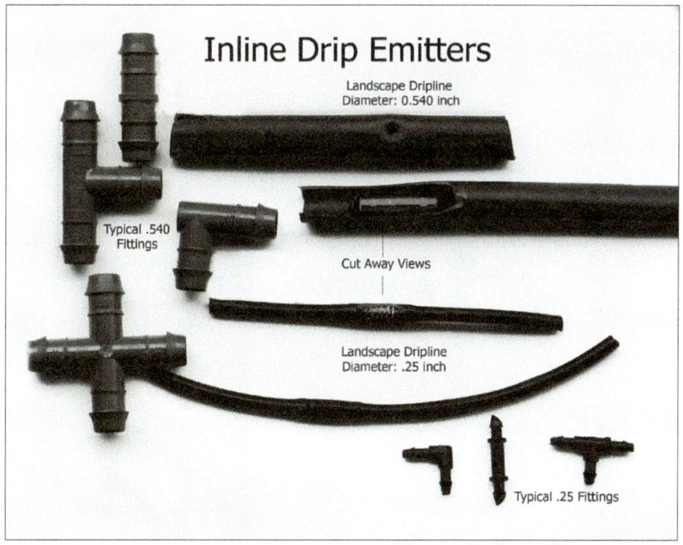

Photo: Graham Hunter

Figure 6-3. Inline drip emitters and tubing.

Photo: Roger Kjelgren

Figure 6-4. Drip irrigation system with inline emitters used along I-15 in north-central Utah.

A fixed-spacing design works best in areas of high plant density like perennial beds.

well in situations where water needs to be delivered at a faster rate or for a shorter time (e.g., large woody plants, some ground covers, and soils with high infiltration rates) (Figure 6-6; Table 6-2).

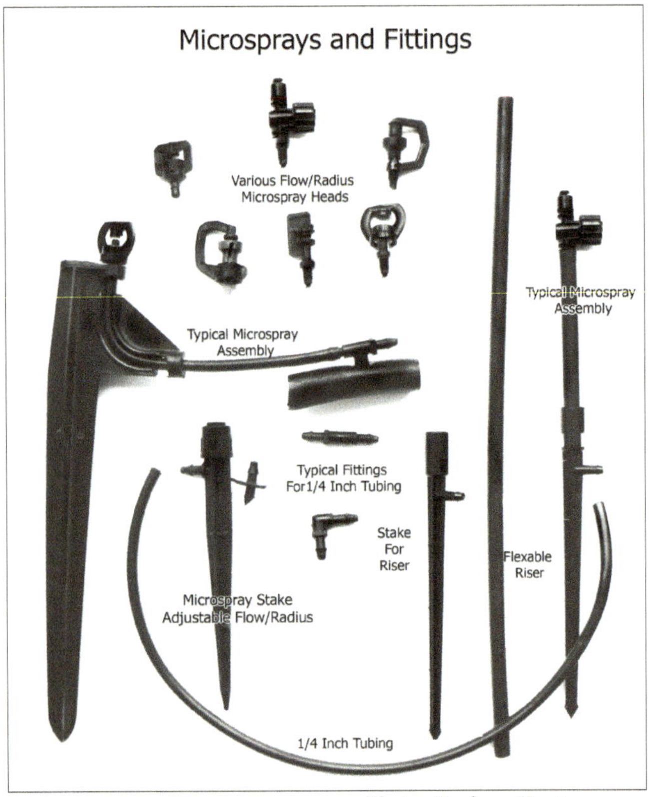

Photo: Graham Hunter

Figure 6-5. Microspray emitters, fittings, and tubing.

Photo: Roger Kjelgren

Figure 6-6. Irrigation system with microspray emitters used in a perennial bed.

Design of Low-Volume Systems

Low-volume irrigation systems have either a fixed- or a random-spacing design. The choice about which to use should be based on the landscape design, and sometimes both will be employed within a single landscape. A fixed-spacing design works well in areas of high plant density such as perennial beds, and is easy to envision, engineer, and install. A planting grid is created using a row spacing that is equal to the spacing of the emitters in the polyethylene tubing. The disadvantage of fixed-spacing design is that everything gets the same amount of water, and some plants may be overwatered.

An alternative to fixed-spacing drip irrigation for high-density plantings is a microsprinkler system with emitters spaced evenly to provide head-to-head coverage. A microsprinkler system also works well for large plants with wide canopies and deep roots, and for low plantings and ground covers in which plant materials are not sensitive to having water on their foliage, and foliage is not so high and dense that it blocks the spray.

Random-spacing designs can accommodate either drip or microsprinkler emitters. In a random-spacing drip system, for example, tubing is snaked through the planting bed and emitters spaced based on the size and position of the plants. This is a good approach for widely spaced plants, and is excellent for plants that do not tolerate water on their leaves (for example, young cliffrose plants). As these plants grow, however, their branches will be held farther from the ground, and microsprinkler emitters can safely be used without harming the plant. (Never use high-pressure sprinkler systems around

Table 6-2. Emitters and Length of Irrigation Time.

PLANT TYPE	SOIL TEXTURE		
	SANDY	LOAM	CLAY
Large trees (canopy 14 feet)	• Microsprinklers; Four or five 25-gph* emitters for 1 hr.	• Microsprinklers; Four or five 15-gph emitters for 1.5 hrs.	• Microsprinklers; Four or five 10-gph emitters for 2.5 hrs.
Medium trees (canopy 10–12 feet)	• Microsprinklers; Two or three 25-gph emitters for 1 hr.	• Microsprinklers; Two or three 15-gph emitters for 1.5 to 2 hrs.	• Microsprinklers; Two or three 10-gph emitters for 2.5 to 3 hrs.
Small trees/large shrubs (canopy 6–8 feet)	• Microsprinklers; Two 20-gph emitters for 0.5 to 1 hr.	• Microsprinklers; Two 10-gph emitters for 1 to 2 hrs.	• Microsprinklers; Two or three 5-gph emitters for 2 to 2.5 hrs.
Medium shrubs (canopy 3–5 feet)	• Microsprinklers; One or two 10-gph emitters for 0.5 to 1 hr.	• Microsprinklers; One or two 5-gph emitters for 1 to 2 hrs.	• Drippers; One or two 2-gph emitters for 2 to 3.5 hrs.
Shrubby ground covers	• Microsprinklers; One or more 5-gph emitters for 0.5 hr.	• Drippers; One or more 1-gph emitters for 2 to 2.5 hrs.	• Drippers; One or more 0.5-gph emitters for 2 to 2.5 hrs.
Herbaceous ground covers	• Microsprinklers; Use microsprinklers spaced evenly to provide head-to-head coverage of the planted area.		
Widely or irregularly spaced perennials	• Drippers; One or two 2-gph emitters per plant; run for 0.5 to 2 hrs.	• Drippers; One or two 1-gph emitters per plant; run for 1 to 3.5 hrs.	• Drippers; One to three 0.5-gph emitters per plant; run for 2 to 4 hrs.
Perennials in a bed	• Set up a low-volume system using either microsprinklers or inline drip emitters laid out in a grid format. Choose lower-gph emitters for clay-type soils and increase the length of irrigation time.		

Tips:

- Plan the irrigation system based on the mature size of plants during the hottest time of the year.
- Check for depth of water penetration using a soil probe (or a metal rod or screwdriver) and adjust irrigation times accordingly.
- In clay soils, plant low-water-use native plants on a berm to keep their roots out of the saturated soil. Cycle irrigation also works to prevent pooling of water at the soil surface.

*gph= gallons per hour

plants with water-sensitive leaves; the spray will inevitably reach the leaves and weaken or kill the plant.)

Whichever system is used, ultimately it should be designed to accommodate the mature size of the plants. Emitter size can be scaled up as plants grow. This way the system will remain functional and deliver the proper amount of water throughout the life of the plants. Remember, roots will not penetrate dry soil. The diameter of the surface wetting zone and depth of wetting must stay just

Never use high-pressure sprinkler systems around plants with water-sensitive leaves; the spray will inevitably reach the leaves and weaken or kill the plant.

The diameter of the surface wetting zone and depth of wetting must stay just ahead of the plant root zone area.

The total flow from emitters should not exceed 75% of the flow rate for the size of tubing. If water coming from a three-quarter-inch line supplies 12 gallons per minute, or 720 gallons per hour, the output that can be supported is approximately 500 gallons per hour. Pressure is lost as the water flows past each consecutive emitter. Pressure loss for each system must be calculated to ensure there is enough pressure to supply each emitter and that enough emitters are employed to avoid "blowing out" the system, or exceeding the capacity of each emitter to handle the pressure. For example, the system described above would provide enough pressure to support twenty-five 20-gph emitters, one hundred 5-gph emitters, or two hundred fifty 2-gph emitters (or some combination of the above).

ahead of the plant root zone area. Trees and shrubs may grow out of a drip-emitter system; in this case, low-volume microsprinker emitters can be substituted for point-source drip emitters with only a few minor adjustments. Some emitters are also adjustable so that the flow can be increased to cover more area as the plant grows. Remember to recalculate pressure tolerance of the system as you scale up the emitter size (see sidebar).

Irrigation Schedule

Scheduling irrigation for a low-water nonturfgrass landscape is more of an art than a science, and many factors influence how often plants should be watered. Time of year, size of plants, weather conditions, and soil texture are just some of the things to consider when deciding how often to irrigate. For example, plants should be irrigated more frequently during the summer months, and irrigation frequency should be adjusted as temperatures cool down. Irrigation system controllers can be programmed with rain sensors that will override the system when a significant rain event occurs. In sandy soils, irrigation frequency may need to be increased, as these soils will dry out faster than water-holding clay soils. In clay soils, water may begin to pool on the soil surface and could run off before the water has reached the depth of the plant's root zone. In this case, slower water application for a longer time may be necessary to meet the water needs of the plant (Tables 6-1 and 6-2). *Cycle irrigation*, watering in multiple short intervals or cycles, may also be used to avoid pooling and runoff of irrigation water in clay soils.

In general, the larger the plant, the more deeply and less frequently it will need to be irrigated (Table 6-3). Advise your clients to pay attention to the plants in their yard. Even the most sophisticated irrigation system cannot take the place of a careful observer who notices when a plant's leaves begin to wilt (often a symptom of underwatering) or turn significantly yellow (often a symptom of overwatering). Symptoms of underwatering can be similar to symptoms of overwatering. Instruct your clients to check their soil periodically between irrigation events. If soil is still wet just prior to a scheduled irrigation, the irrigation schedule may need to be adjusted.

Some clients may feel uncomfortable with the technology of automatic irrigation systems. In this case, there is nothing wrong with dragging a hose around to individual plants. In fact, a garden based largely on very low-water-use plants will have very little need of an expensive irrigation system when all that is required is a deep watering once or twice per month during the summer. In fact, the more drought-tolerant the plants, the less precise the schedule can be. When water runs short, drought-tolerant plants are unlikely to die; they just grow more slowly. Even so, it is important that all plants be watered regularly for the first season or two until they establish an extensive root system.

Table 6-3. How often and how deep to water by plant type and rooting depth.

IRRIGATION SCHEDULE		DEEP WATERING MEANS:
Very low-water-use plants	Deep watering once or twice during the growing season until established	*Tree roots* 18 to 24 inches
Low-water-use plants	Deep watering every 4 to 6 weeks during the growing season	*Shrub roots* 12 to 18 inches
Medium-water-use plants	Deep watering every 2 to 4 weeks during the growing season	*Herbaceous plant roots* 6 to 12 inches
High-water-use plants (turfgrass or annual beds)	Deep watering 2 or 3 times per week during the growing season	*Turfgrass roots* Up to 6 inches

Tips:

- Irrigate less frequently during cool weather and more frequently during hot weather.
- Irrigate early in the day to avoid water loss due to evaporation.
- Apply mulch around plantings to minimize evaporation.
- Watch plants for signs of over- or underwatering and adjust the schedule accordingly.

Irrigation for Turfgrass and Other High-Water-Use Plants

Higher-pressure sprinkler irrigation systems are best suited for relatively large areas (greater than 10 × 10 ft.) with a uniform surface planting, such as turfgrass or other type of ground cover. The sprinkler emitters should be spaced to provide head-to-head coverage and their spray pattern adjusted to water only planted areas.

Advantages of a conventional sprinkler system are its relatively low installation cost, low maintenance, and decreased risk of damage. High-pressure sprinkler systems can operate with less filtering due to the large size of the nozzles in the spray head. A disadvantage is the increased risk of watering outside the plant root zone, which can waste water and encourage weed germination and growth. Higher operating pressures can result in more waste due to overspray, especially in windy areas. With more moving parts and higher pressures, high-pressure sprinkler systems can be vulnerable to breakdown from the constant pounding of the water. Water conservation with sprinkler irrigation can be realized by proper scheduling.

Sprinkler Irrigation Scheduling

Irrigation scheduling is perhaps the pinnacle of turfgrass irrigation system management. One could have a perfectly designed system, but still waste water if the site manager does not know how to properly schedule irrigations. Once installed, irrigation systems should have an *irrigation water audit.* Irrigation water audits enable the site manager to determine the rate and evenness of water application to plants, and to correctly schedule irrigations.

Irrigation Water Audits

The first step in water auditing is to conduct a quick run-through of the irrigation system. With one zone running at a time, check to see that all sprinkler heads are in their best possible alignment, throwing water head-to-head and not on pavement. Also check for broken sprinkler heads or anything else that may be impeding the uniformity of irrigation, such as boulders or sunken sprinkler heads. Once all zones have been inspected, a catch cup test should be conducted (see the Catch Cup Test Worksheet at the end of this chapter). Catch

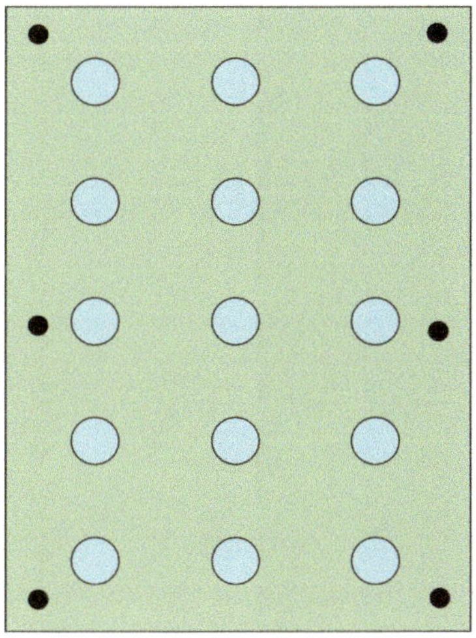

Figure 6-7. Single zone layout of catch cups.

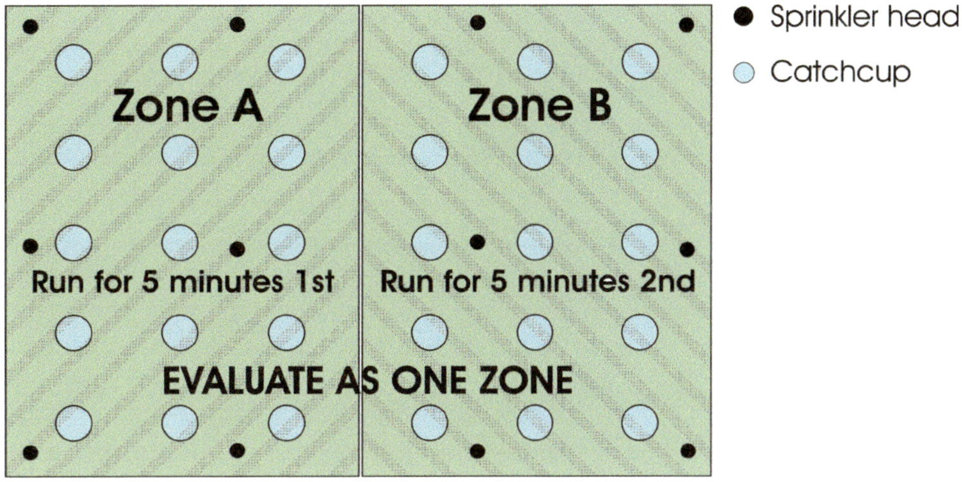

Figure 6-8. Layout of catch cups for overlapping zones.

cups specifically designed for irrigation system audits are available in the industry, but any straight-sided container will do. To obtain catch cups from the Utah State University Center for Water-Efficient Landscaping, call 435-797-2255 or 435-797-1523.

Catch cups set out in a uniform pattern within an irrigation zone will enable you to determine the application rate of irrigation in that zone. If zones happen to overlap, then set up catch cups in both adjoining zones and run the zones sequentially. The application rate will apply to both of those zones combined. The pattern in which the cups are laid is not as important as even spacing. A good rule of thumb is to set the cups *at* a head and *between* two heads, as shown in Figure 6-8. It is important to offset the cups enough that they are not impeding the sprinkler pattern. Offsetting cups by 12 inches from the sprinkler head will usually suffice for most systems, but the distance may need to be increased with larger, lower-angle rotary heads.

With the cups laid out in a pattern, run the zone for approximately 10 minutes, or until you can get enough water in the catch cup to make an accurate measurement. Record the number of minutes the zone was run. Next, record how much water is in each catch cup. It is sometimes helpful to draw a map of the zone, and record where the catch cups are and how much water they contain. This is a good tool for spotting problem areas within zones.

Once each catch cup water level is recorded, total all of the measurements and divide by the number of catch cups. For example, if the total inches of water in the catch cups is 2.4, and there were 12 catch cups in the test, *the average water level over all the cups is 0.2 inches.*

Total water level of all cups (inches) ÷ number of catch cups = average water level of catch cups (inches)

Next, find the average water level in the 25% of cups that contain the lowest amounts of water. First, determine how many cups constitute 25% of the total cups. If there were 12 catch cups in the test, multiply this number by 25% to get the number of low-amount cups to use. In this case, the number of low-amount cups to use is three.

Number of catch cups × 0.25 = number of low-amount cups to use

Add the inches of water in each of the three cups that contain the lowest amount, and divide that by the total number of cups. For instance, if the water level in the lowest three cups is 0.09, 0.11, and 0.11 inches, the total amount of water in those three cups is 0.31 inches. Divide 0.31 by three to get 0.1. *This is the average water level in the 25% of catch-cups with the lowest amount of water.*

Total water level in lowest cups (inches) ÷ number of low-amount cups = average water level of the lowest-water-level cups (inches)

After finding the average water level of the low-amount cups, divide this number by the average water level of all cups. This number represents the uniformity, or evenness, of the system being audited. In this case, we would divide 0.1 inch by 0.2 inch to get a 50% uniformity of water application.

Average water level of low-amount cups (inches) ÷ average water level of all cups (inches) × 100 = uniformity of application (%)

The uniformity of application percentage is an important tool. It allows you to evaluate the condition of your system by comparing it against an established standard. While no irrigation system is perfect, there are some generalized recommendations for different kinds of systems. For systems with pop-up spray heads, a realistic uniformity rating is 65%, while the standard for rotary spray-head zones is 70%. If the system uniformity indicated by your irrigation system audit is lower than the standard, extra time should be devoted to evaluating the system inefficiencies (see Irrigation System Maintenance at the end of this chapter).

Calculating Irrigation Timing and Cycling

The general recommendation for the amount of water to use per irrigation event on turfgrass in the Intermountain West is one-half inch (or five-eighths inch in the southern part of the region). Applying this amount of water every time you irrigate will simplify irrigation scheduling. The only factor in the irrigation schedule that will change will be the number of days between irrigation events.

To determine how many minutes will be required to apply one-half inch of water to the zone, multiply the number of inches you wish to apply (one-half inch) by the number of minutes the system was run in your audit. Then, divide that number by the average water level of all catch cups used in your audit. Using the previous sample audit: one-half inch of water multiplied by 10 minutes system run-time equals five; five divided by 0.2 (average water level of catch cups after 10 minutes) equals 25. You will need to run the irrigation system for 25 minutes to apply one-half inch of water to that zone. Repeat this procedure for each zone in your system.

Number of inches to apply × number of minutes used in audit ÷ average water level over all catch cups (inches) = number of minutes to apply desired amount of water

Once the number of minutes to run the zone has been determined, the number of cycles within the irrigation event needs to be determined. This is especially important in zones with heavy soils or significant slopes. Turn on the irrigation system and record how long it takes for you to see water pooling or running off the landscape. This is the maximum number of minutes the zone can be run at one time. To determine how many cycles are appropriate for the specific zone, divide the total minutes the system will run by the number of minutes before runoff occurred. In this case, imagine five minutes passed before runoff was observed. Dividing that into 25 minutes gives the number of cycles to run, which in this case is five.

Total minutes to apply desired amount of water ÷ number of minutes until runoff = number of cycles to schedule

Now that you have the information you need to schedule the timing and cycling of each irrigation event, it is time to determine the number of days between these events.

Water Audit Checklist

- [ ] Turn the system on and check each zone independently for problems with broken, blocked, misaligned, or sunken spray heads.
- [ ] Conduct a catch cup test:
- [ ] Arrange catch cups in a uniform pattern, evenly spaced within an irrigation zone.
- [ ] Run the irrigation zone for about 10 minutes. Record the exact number of minutes.
- [ ] Record the number of inches in each catch cup.
- [ ] Total all measurements and divide by the number of catch cups.
- [ ] Find the average water level in the 25% of cups that contain the lowest amounts of water.
- [ ] Divide the average water level of the low-amount cups by the average water level of all cups and multiply by 100. This represents the system uniformity.
    - Pop-up spray heads: expect up to 65% uniformity.
    - Rotary heads: expect up to 70% uniformity.
- [ ] Apply one-half to three-quarters inch of water to turfgrass each time you irrigate:
    - Number of inches to apply × number of minutes used in audit ÷ average inches of water over all catch cups = total run time
- [ ] For cycle irrigation, turn on the irrigation system and record how long it takes to see water pooling or running off the landscape. Divide the total run time by the number of minutes to pooling or runoff to get the number of cycles needed.

Evapotranspiration Method for Calculating Water Needs

Evapotranspiration (ET) is the term irrigation managers use to determine how much water is needed to replace what is lost from their landscapes. Evapotranspiration is the loss of water from the soil both by evaporation and by transpiration through plants. ET is determined by a number of factors, including solar radiation, relative humidity, wind speed, and temperature. ET is recorded as a rate, such as inches per day, and is calculated from data measured by a weather station.

Weather stations can be used to determine a very accurate irrigation schedule based on real-time data; however, many sites do not have the luxury of an accessible weather station. In Utah and some other intermountain western states, ET information has been made available online. The Utah Division of Water Resources provides this information as a part of its ongoing water conservation program. Data can be found online at http://www.conservewater.utah.gov.

Several agencies in Utah, including Utah State University Extension, the U.S. Bureau of Reclamation, and the Utah Division of Water Resources have developed a generalized watering schedule based on historic, 30-year-average ET data. These data, when used with the irrigation scheduling program described above, provide the following watering schedules for northern and southern Utah:

Photo: Graham Hunter

Figure 6-9. Weather station at Greenville Farm, North Logan, UT.

Table 6-4. Irrigation schedule for northern Utah.

NORTHERN UTAH	APPLY ONE-HALF INCH OF IRRIGATION
Startup until April 30	Once every 6 days
May	Once every 4 days
June	Once every 3 days
July	Once every 3 days
August	Once every 3 days
September	Once every 6 days
October 1 to Shutdown	Once every 10 days

Table 6-5. Irrigation schedule for southern Utah.

SOUTHERN UTAH	APPLY FIVE-EIGTHS INCH OF IRRIGATION
Startup until March 31	Once every 8 days
April	Once every 5 days
May	Once every 4 days
June	Once every 3 days
July	Once every 3 days
August	Once every 3 days
September	Once every 5 days
October	Once every 7 days
November 1 to Shutdown	Once every 10 days

Irrigation System Maintenance

Irrigation system maintenance is necessary to ensure the most efficient use of water. The following maintenance recommendations are designed to help with irrigation system evaluation.

Irrigation Controller

Irrigation controllers should be checked at the beginning of each growing season, before running sprinklers for the first time. To do this, open the controller cabinet and clean out cobwebs, dirt, and debris. Next, check the time and day displayed on the controller, and correct them if necessary. This is also a good time to change the battery and check wiring for loose connections. Check all wire connections, including the rain sensor connection if one is attached. Consider adding a rain sensor to the controller if one is not already included. A rain sensor is inexpensive, simple to install, and will automatically shut off the irrigation system when a significant amount of rain falls.

Low-Volume Systems

Low-volume irrigation systems have a great advantage in water savings over conventional sprinkler systems, but they do require more attention to keep them running efficiently. Disadvantages of a drip system include higher installation costs and a higher risk of damage to aboveground and shallowly buried parts. Possibly the biggest disadvantage is the need to rethink irrigation and general landscape maintenance practices. One careless person with a dandelion digger can lead to a whole lot of time repairing a perforated system. A valve with eight spray heads in a conventional sprinkler irrigation system requires little more than a glance to troubleshoot problems; a valve in a low-volume system with a thousand buried inline emitters requires close scrutiny and, most of all, a careful, observant operator.

To maintain a low-volume irrigation system, flush the system for a few minutes with an open flush valve before running the system for the first time each season to remove dirt and debris. Close the flush valve and run the system, one control valve at a time, to check for problems.

- Clogged emitters should be replaced. *Install a water filter if the system does not already have one.*
- Check the placement of emitters. Emitters should be at the edge of the root ball of new plants and moved to the drip line (edge of the canopy) of established plants.
- Check for emitters in which tubing has popped off due to high pressure; install a pressure regulator if needed.
- Check to see that all emitters are in place. Missing and broken emitters should be replaced to keep the system running efficiently.
- Look for pinched or broken tubing and straighten or replace it. Make sure that all tubing is attached to the appropriate emitters, and that connections are secure. Make necessary adjustments and repairs to the system. When this is done, turn the irrigation system on manually to make sure it is operating as programmed.

Sprinkler Systems

Once the controller has been programmed with the proper irrigation schedule, inspect the sprinkler system by checking valves, sprinkler heads, and emitters. Before running the system for the first time, remove the last sprinkler head in each line and let the water run for a few minutes to flush out dirt and debris. Replace the sprinkler head and turn the system on.

- Evaluate spray patterns and positions of sprinklers for obvious problems such as clogged or misaligned heads.
- Some sprinkler heads may be tilted, obstructed by grass, or even buried. If not positioned properly, these sprinkler heads will be unable to apply water efficiently.
- Some sprinklers have built-in filter screens that should be cleaned and replaced, if necessary.
- Leaks and misting from sprinkler heads may indicate high-water-pressure problems. High-pressure problems can be corrected by plumbing a

pressure regulator into the sprinkler system. Pressure regulating sprinkler heads are also available.

- Make necessary adjustments and repairs to the system in order to apply the water as evenly as possible. The flow control on the valves may be adjusted to fine-tune the system. When this is done, turn the irrigation system on manually to make sure it is operating as programmed.

Winterization

Basic irrigation system winterization is simple. The water supply should be turned off at the main valve, and the irrigation controller should be set to the "rain" or "off" setting. After doing this, each valve should be turned on to release pressure in the pipes, and water should be drained from the system to protect components that could freeze. The system may have drain valves that can be opened for drainage, or the system can be blown out using air.

The goal of irrigation system maintenance is to create the most efficient irrigation system possible so that water is not wasted on the landscape. While perfect efficiency is impossible to achieve, most irrigation systems can be dramatically improved by regularly following these simple maintenance practices. Irrigation systems should be examined on a regular basis (each spring and several times during the growing season), to keep them operating at peak efficiency. Most importantly, use an irrigation schedule that accounts for plants' changing needs over the growing season.

CATCH CUP TEST WORKSHEET

Name:________________________________	Address:________________________________
Date:________________________________	Catch Cup Test By:________________________
Test 1: Zone_____	Test 2: Zone_____
Head Type: Spray Rotor Pressure: _____ psi	Head Type: Spray Rotor Pressure: _____ psi
Soil Type: Clay / Loam / Sandy	Soil Type: Clay / Loam / Sandy
Root Depth: ______ inches	Root Depth: ______ inches

	Cup Depths (cm/in)		Cup Depths (cm/in)		Cup Depths (cm/in)		Cup Depths (cm/in)
1.	______	17.	______	1.	______	17.	______
2.	______	18.	______	2.	______	18.	______
3.	______	19.	______	3.	______	19.	______
4.	______	20.	______	4.	______	20.	______
5.	______	21.	______	5.	______	21.	______
6.	______	22.	______	6.	______	22.	______
7.	______	23.	______	7.	______	23.	______
8.	______	24.	______	8.	______	24.	______
9.	______	25.	______	9.	______	25.	______
10.	______	26.	______	10.	______	26.	______
11.	______	27.	______	11.	______	27.	______
12.	______	28.	______	12.	______	28.	______
13.	______	29.	______	13.	______	29.	______
14.	______	30.	______	14.	______	30.	______
15.	______	31.	______	15.	______	31.	______
16.	______	32.	______	16.	______	32.	______

Test 1	Test 2
Total=______________	Total=______________
# of Cups=__________	# of Cups=__________
Time=______________	Time=______________

Test 1 Calculations:

Precipitation Rate

_____ + (_____ / 60 = _____) = ________
Minutes Seconds Decimal Total Time

________ / _____ = ________
Cup Depth Total # of Cups Cup Average

_______ / ________ x 60 = ____________
Cup Average Time as Decimal Precipitation Rate (in./hr.)
(x .3937 if measured in cm)

Distribution Uniformity (DU)

Find the lowest 25% of the depths above.

______ x 0.25 = ______ (find this many cups...)
Total # of Cups 25% of Cups

1. ______ 4. ______ 7. ______
2. ______ 5. ______ 8. ______
3. ______ 6. ______ 9. ______

= _________ / ________ = ________
Sum of Low Cups 25% of Cups Low Average

________ / _______ = ___________ %
Low Average Total Average DU

Test 2 Calculations:

Precipitation Rate

_____ + (_____ / 60 = _____) = ________
Minutes Seconds Decimal Total Time

________ / _____ = ________
Cup Depth Total # of Cups Total Average

_______ / ________ x 60 = ____________
Total Average Time as Decimal Precipitation Rate (in./hr.)
(x .3937 if measured in cm)

Distribution Uniformity (DU)

Find the lowest 25% of the depths above.

______ x 0.25 = ______ (find this many cups...)
Total # of Cups 25% of Cups

1. ______ 4. ______ 7. ______
2. ______ 5. ______ 8. ______
3. ______ 6. ______ 9. ______

= _________ / ________ = ________
Sum of Low Cups 25% of Cups Low Average

________ / _______ = ___________ %
Low Average Total Average DU

Irrigation System Maintenance Checklist (Drip Irrigation)

Start of the Season:

- ❑ Flush the system for a few minutes with an open flush valve. Close the flush valve and run the system, one control valve at a time, to check for problems.
- ❑ Replace clogged emitters.
- ❑ Check the placement of emitters. (Emitters should be moved to the drip line or edge of the canopy of established plants.)
- ❑ Check for emitters in which tubing has popped off due to high pressure. (You may need to install a pressure regulator if this becomes a persistent problem.)
- ❑ Check to see that all emitters are in place. Missing and broken emitters should be replaced to keep the system running efficiently.
- ❑ Look for pinched or broken tubing and straighten or replace it.
- ❑ Turn the irrigation system on manually to make sure it is operating as programmed.

End of the Season:

- ❑ Turn water off at the main valve.
- ❑ Set the irrigation controller to the "rain" or "off" setting.
- ❑ Turn on individual valves to release pressure in the pipes and drain water from the system.

Irrigation System Maintenance Checklist (Sprinkler Irrigation)

Start of the Season:

- ❑ Remove the last sprinkler head in each line and let the water run for a few minutes to flush out dirt and debris.
- ❑ Replace the sprinkler head and turn the system on.
- ❑ Evaluate spray patterns and positions of sprinklers for obvious problems such as clogged or misaligned heads.
- ❑ Clean and replace filter screens if needed.
- ❑ Look for leaks and misting from sprinkler heads. These may indicate high water pressure problems.
- ❑ Adjust flow control on the valves to fine-tune the system to apply water evenly.
- ❑ Turn the irrigation system on manually to make sure it is operating as programmed.

End of the Season:

- ❑ Turn water off at the main valve.
- ❑ Set the irrigation controller to the "rain" or "off" setting.
- ❑ Turn on individual valves to release pressure in the pipes and drain water from the system.

Principle 7

Landscape Maintenance

Amy Croft, Larry A. Rupp, and Heidi A. Kratsch

Maintenance techniques for a water-efficient landscape are not different from those for a conventional landscape. But differences do exist regarding the types and frequencies of some tasks. Due to the predominance of turf, the maintenance for conventional landscapes usually involves frequent mowing and irrigation throughout the growing season. In a water-efficient landscape, the task of mowing may be reduced, but it is often replaced by weeding, deadheading, and mulching. Most water-efficient landscapes will require irrigation, although water-efficient gardens dominated by plants with very low water requirements may need only infrequent irrigation during the hottest part of the year. Both conventional and water-efficient landscapes require pruning, fertilizing, pest management, and other activities on an intermittent basis.

It is important to realize that a water-efficient landscape is not a maintenance-free landscape.

It is important to realize that a water-efficient landscape is not a maintenance-free landscape. Similar to a conventional landscape, a water-efficient landscape will need more maintenance time for the first couple of years due to the weed control and mulching required while waiting for perennials, trees, and shrubs to mature and fill in the open spots in planting beds. As plants mature, they will begin to outcompete the weeds, thus reducing maintenance time. The main elements involved in water-efficient landscape maintenance are irrigation, weed control, fertilization, pruning, observation, and insect and disease control. Since water-efficient landscapes are designed to operate on a minimum amount of external inputs such as water and fertilizer, it is even more important to carefully observe them and ensure that they continue to get the required care for desired performance. In addition to problems caused by lack of maintenance, water-efficient landscapes are often seriously compromised by overirrigation, with subsequent consequences to the health of drought-adapted plants.

Irrigation

While closure of plant canopies can reduce weed infestations, it can also change the irrigation requirements of the landscape. An obvious characteristic of any desert is that plants are spaced much farther apart than they would be in a riparian area or other situation with adequate water. Such spacing is a function of the available water and is determined by plant competition. If water is more limited, plants will be smaller and farther apart; if water is less limited, plants will be larger and closer together. Irrigation management in a water-efficient landscape should take into account the higher water requirements of plants as they become larger and fill in more of the landscape space. Therefore, managing the density and size of plantings is an important means of managing irrigation requirements.

Proper irrigation is essential to maintaining a water-efficient landscape. The amount of water applied to a landscape will affect all other aspects of maintenance, and is critical to a successful planting. All plants will require supplemental water until they have established an adequate root system, but irrigation of drought-tolerant plants can be cut back after that time, usually after one season for herbaceous perennials and after one to two years for trees and shrubs. Irrigation frequency, especially for trees, should be reduced gradually to allow plants to redevelop roots lost during transplanting. A return to a normal rate of twig growth is a good indication that the tree is beginning to put on new roots. The frequency of irrigation can also be used to affect the rate at which new plantings are brought to their desired size; reducing irrigation frequency when plants have well-established roots can reduce the amount of plant waste produced and, therefore, reduce time spent mowing and pruning established plants. In essence, there is one irrigation protocol for growing new landscape plants, and a different protocol for maintaining them. Once plants are established, deep watering on a less frequent basis is more beneficial than frequent, shallow watering. Intermittent deep irrigation encourages plant roots to grow deeper and wider, thus enhancing their drought resistance. Frequent, shallow watering limits root growth and can weaken or kill some drought-tolerant species.

It is important to realize that more water-efficient landscapes fail because of overwatering than they do from underwatering. Care must be taken to ensure plants are given the proper amount of water. This can be a difficult task because many trees, shrubs, and perennials do not show water stress as readily as turfgrass.

More water-efficient landscapes fail because of overwatering than they do from underwatering.

Techniques such as observing indicator plants (plants that readily wilt under water stress), using soil moisture sensors, or using weather-based irrigation controllers can help to keep irrigation at optimal levels. Using minimum irrigation rates and watering plants early in the day help prevent weeds, insect pests, and disease-causing microbes from invading the water-efficient landscape. An important caveat is that when water is reduced to the point of stressing the plant, some diseases or insects may actually become worse due to the plant being in a weakened state. It is critical to be observant and knowledgeable about plant stress symptoms to balance water conservation needs with the health and aesthetic requirements of landscape plants.

From a practical point of view, *a huge part of landscape maintenance is keeping the irrigation system working efficiently*. This is even more critical in a water-efficient landscape where plants are irrigated to meet minimum water requirements. A broken irrigation system can quickly lead to drought stress in one part of the irrigation zone, and possibly to deadly flooding of sensitive plants in another part of the irrigation zone. The time required for repairing broken spray heads or emitters, fixing leaks, trimming plants that block water spray, cleaning filters on pumps and spray heads, unplugging drip emitters, and fixing tubing damaged by rodents can be significant. These challenges are even greater in a water-efficient landscape in which a drip irrigation system is used, and where emitters are buried by soil or mulch; the only indication of an irrigation problem may be when the affected plants die. A regular schedule of observation and irrigation system maintenance should be established (see "Principle 6: Irrigation Planning" for irrigation system maintenance checklists).

The choice of irrigation method can also affect weed germination and growth. Infrequent watering reduces weed seed germination, but it also influences which weed species predominate in a landscape. Drip irrigation helps reduce weeds by applying water only where it is needed by desirable plants. Soil between plants is left dry and less favorable for weed germination and growth. Where weeds do establish, the mix of weed species is altered. Typical weeds found in well-watered gardens, such as pigweed and lambsquarters, may be replaced by more drought-tolerant weeds such as Russian thistle and kochia. As the population of weed species changes, it may be necessary to change weed management practices.

Weed Control

Weed control is essential to a successful landscape. The first step to weed control is to ensure that plants or soils brought into the landscape are free from weeds.

Weed control is essential to a successful landscape.

This is especially true for perennial weeds such as bindweed or quackgrass, which can be very difficult to remove from the landscape once they are established. It can also be difficult to determine whether or not topsoil is free from weed seeds. Often the best that can be done is to know the source of the soil and use that information to infer what weed problems may exist. It is possible to inspect soil for vegetative weed propagules such as rhizomes.

New plants should be inspected to ensure that they are in good overall health and that there are no weeds present. In the event there are uncontrolled weeds in the landscape, a healthy plant is more likely to outcompete them for limited water and nutrients. If weeds do invade, there is much less impact to the landscape in terms of soil disturbance and waste material if the weeds are removed when they are as small as possible. It requires less labor to remove a small weed than it does to extract a large, well-rooted one. Usually, by staying on top of weed problems throughout the growing season, a thorough weeding in the spring will be all the maintenance that is required. Never allow weeds to go to seed or they will continue to be a problem in subsequent seasons.

Mulching is a key element of weed control in water-efficient landscapes. A thick layer of mulch (two to three inches) will prevent light from reaching weed seeds so they cannot germinate. Weeds that do germinate in a mulch bed are easier to pull than they would be in unmulched soil. Mulches generally consist of organic materials such as shredded or chipped wood or bark, or of inorganic materials such as gravel or rock. Organic mulches require annual replenishment, which adds to the maintenance requirement. Rock mulches are more permanent; but they can accumulate organic debris from landscape plants, and they are harsher on the landscape environment because they retain heat in the root zone and do not contribute to improved soil structure (see "Principle 5: Use of Mulch" for more information).

If weed barriers are used under mulch, they should be porous woven fabric, not smooth plastic films. Weed barriers tend to work best when applied under rock mulches where they can reduce weeds and prevent the mixing of rock and soil. Weed barriers under organic mulches are less effective because they prevent the addition of organic matter from decomposing mulches to the soil, they allow weed germination in decomposed mulch on top of the soil, and they encourage moist soils with shallow root growth. Mulches can be an excellent tool in helping to retain soil moisture; however, they can hide irrigation problems. For example, a broken spray head that saturates the soil beneath mulch might never be noticed until serious damage has occurred.

Where hand-weeding, mulches, and competition from desirable plants fail to control weeds, another option is the use of herbicides. Herbicides are designed to control weeds through various modes of action such as selectivity (2,4-D to remove dandelions from grass), pre-emergence (trifluralin to prevent germination of weed seeds in a perennial or shrub bed), or directed foliar application (directed sprays of nonselective post-emergence herbicides such as glyphosate on perennial weeds). It is important to understand the types of herbicides, their recommended uses, and the plants susceptible to them in order to have an effective weed control program that does not damage desirable plants. Using herbicides in mixed plantings of perennial flowers, shrubs, and trees is more difficult than in monoculture landscapes such as athletic fields, because some nontarget plants may be injured. Always follow the label directions and manufacturer's application recommendations when using herbicides. Keep in mind, if valuable plants are damaged by herbicides used outside label recommendations, the applicator bears the liability.

It is important to understand the types of herbicides, their recommended uses, and the plants susceptible to them in order to have an effective weed control program that does not damage desirable plants.

Timely use of pre-emergent herbicides and directed sprays of nonselective materials are the most useful for water-efficient landscapes. Selective herbicides are difficult to use in mixed landscapes. While they may be safe for one plant, a different plant immediately adjacent may be susceptible. The only safe use for selective herbicides is to remove broadleaf weeds from turf areas, or to remove grassy weeds from broadleaf plantings such as shrub beds. Directed sprays of herbicides such as glyphosate are very effective in removing perennial weeds, as long as the spray is limited to contact with the weed. With many perennial weeds, cultivation (disturbance of the soil) actually worsens the weed problem by increasing the number of propagules; therefore, directed sprays may be a wiser approach.

In water-efficient landscapes in which many perennial plants are used, pre-emergent herbicides can be an effective means of reducing weed seed germination. An example is the use of trifluralin to control cheatgrass in juniper beds. The use of pre-emergent herbicides in an area does eliminate the option for growing annual plants from seed or for reseeding perennial flowers in that area.

Fertilization

Plants adapted to dry soils are more likely to be adapted to low-nutrient environments. This adaptation makes fertilizing a water-efficient garden an easy task because the dry zones should need little if any fertilization. The only time fertilization may be necessary is during the establishment years when stimulating

One way to increase professionalism and to ensure proper use of fertilizers is to apply them based on soil test recommendations.

rapid growth will achieve a more desirable appearance. Adding excess fertilizer and moisture to plants that are adapted to dry, low-fertility soils adds to maintenance time by stimulating excess growth and creating large and unruly plants; it may also weaken or kill sensitive plants. Do not prune or fertilize when water is scarce, as the newly stimulated growth will be faced with water stress. Plants growing in hydrozones requiring more water may need more fertilizer than others, because nutrients are depleted from the soil more rapidly. Given the potential for plant damage, the high cost of fertilizer, and the concerns about excess fertilizers polluting the environment, it is important to fertilize only as necessary. One way to increase professionalism and to ensure proper use of fertilizers is to apply them based on soil test recommendations. There are soil testing labs at many land grant universities, and tests may be provided by private companies. These labs analyze soil samples and provide recommendations regarding how much fertilizer should be applied for optimal plant growth in a given situation. In Utah, samples may be sent to the Utah State University Analytical Laboratory (http://www.usual.usu.edu) via any county extension office.

Insect and Disease Control

The climate of the Intermountain West is generally described as a high desert. This means most of our annual moisture comes in the form of snowpack in the winter, while summers are dry. By using plants adapted to our climate, we limit the chances that pests and diseases will invade the garden. Adapted plants include introduced plants such as lilac or iris, as well as plants native to the region. It is critical to remember, however, that some native plants, such as quaking aspen, are not adapted to growing in our valley environments, and are very prone to a number of pest problems. Unless the goal of the garden is an entirely native landscape, it is best to evaluate each plant on its own merits and determine if it is adapted to the landscape situation. Plants adapted to the water, nutrients, temperatures, and light in a landscape are less likely to fall prey to common pests and diseases because they thrive under the conditions provided. In contrast, plants that are not adapted to local conditions will be under stress and therefore more vulnerable to pests.

In addition to choosing plants that are adapted to the regional climate, it is wise to choose plants that have been acclimated to local landscape conditions. Even tough, native plants can be damaged if removed from a warm greenhouse

Plants adapted to the water, nutrients, temperatures, and light in a landscape are less likely to fall prey to common pests and diseases because they thrive under the conditions provided. In contrast, plants that are not adapted to local conditions will be under stress and therefore more vulnerable to pests.

and immediately placed in a harsh environment. Things to consider in plant selection are the cost of transportation, diversity to avoid monocultures, and resistance to local pest and disease problems. Should pests become a problem, Integrated Pest Management (IPM) techniques are a good first step to controlling them. IPM takes advantage of all available methods of managing pests including the use of pest-resistant plants, cultural practices, biological controls, and pesticides. An important concept in IPM is monitoring to determine when pest levels have reached a predetermined damage threshold *before control is instituted*; a few insects causing only minor damage can provide a critical component of habitat for native birds and other wildlife. More information on IPM can be found at http://utahpests.usu.edu/. This site also links to the Utah Plant Pest Diagnostic Laboratory, which is a very valuable resource in identifying disease or insect problems.

Pesticides are usually developed for use in crops of high economic value such as corn or soybeans. Occasionally, those pesticides are shown to have value in specialty crops such as landscape plants. Because of the large number of species and cultivars used in landscaping, there is a very good chance that within any water-efficient landscape there will be plants that have not been tested for use with a specific herbicide, insecticide, or fungicide. Therefore, it is important for the applicator to use pesticides according to label requirements to ensure that no damage occurs to desirable plants. Extra caution may be required when working with new cultivars or other unknown plants.

In order to determine which pesticides are options for controlling a particular disease or insect, the best approach is to first positively identify the insect or pathogen involved. Once the pest is identified, information on control, including recommended pesticides, can be found in many Utah State University Extension fact sheets. By law, recommendations for application rates of pesticides must be included on product labels, and are not provided in university publications.

Pruning

Another important aspect of water-efficient garden maintenance is proper pruning. Pruning has a number of purposes including promotion of healthy plant structure for woody plants, removal of spent blossoms or seeds, shaping of plants to increase aesthetic value, and control of plant growth to facilitate access and safety in the landscape. Pruning is an effective tool for both introduced and native plants. The pruning techniques used are the same for both and should be based on landscape goals, regardless of whether the plants are native. In addition to water conservation, some water-efficient landscapes are designed to provide a naturalized appearance. In such landscapes, pruning methods such as shearing for highly structured hedges or topiary are not appropriate. On occasion, plants may need to be trimmed to allow free movement along a path or to increase visibility near streets, but a better approach is to plan spacing for the mature size of plants at the time of planting. The goal should be to retain the natural appearance rather than to shear the plants. A good rule of thumb for pruning plants for a natural look is to ensure that when pruning is finished, no one is able to tell the plant has been pruned. All pruning of woody plants should be done with the health of the plant in mind and should follow techniques aimed to increase compartmentalization and prevent invasion by disease-causing organisms; these techniques will minimize plant stress. Pruning should also be done to increase plant structural integrity to avoid damage from storms or other potential problems.

Pruning can be done at any time of the year but is generally done in the late winter or early spring to ensure optimal plant growth. It is also easier to see plant structural problems when branches are bare. However, there are periods (such as just after new shoots have formed in the spring) when pruning can severely set back plant growth. From a management perspective, the opportunity to prune during the winter provides an excellent way to extend the season for plant maintenance activities, and can be used to keep employees busy during the off-season. There are some rules to the timing of pruning specific plants; for example, spring-flowering shrubs should be pruned just after bloom, and maples should be pruned early enough in the winter that they do not "bleed" sap when cut.

Most trees and shrubs benefit from *thinning* cuts that open up their canopy and eliminate old or competing stems. Thinning cuts are made by cutting a branch back to its point of origin. The point of origin could be another branch or the main trunk, or it could be at the base of the plant in some shrubs. An example of a plant thinned to the base is red-twig dogwood, in which old canes are removed annually to make room for the colorful new bright red growth. Thinning can be used to shape, direct growth, or reduce the size of a plant. No more than one third of the old canes should be removed in a growing season. A *heading* cut is more severe than a thinning cut, and removes part of a branch leaving a short stub above a bud. Heading cuts are used to fill in a gap in the canopy or to increase flower production by stimulating a flush of growth from a lateral bud or buds just below the cut. Sometimes heading cuts are mistakenly used when thinning cuts would have been more appropriate. Overuse of heading cuts can ruin the natural shape of a tree or shrub.

If the concept of water-efficient landscaping is expanded to embrace sustainable landscaping, then the overall landscape plan should address ways to reduce labor, fertilizer, fuels, and pesticides in addition to reducing irrigation requirements.

Some shrubs and many herbaceous perennials benefit from occasional *rejuvenation pruning*, which involves shearing the entire plant back close to its base, without cutting into the *crown* (the growing point). This technique can be used on some plants that have become leggy or are beginning to "flop." Examples of plants that benefit from periodic rejuvenation pruning include Blue Mist spirea, fernbush, Apache plume, and rabbitbrush. Not all plants will survive this severe form of pruning; use it only for plants known to respond well. For more information on pruning specific Intermountain West native and non-native plants, see the Utah State University Extension publication "Water-Wise Landscaping: Plant Maintenance," http://extension.usu.edu/files/publications/publication/HG_Landscaping_2008-01pr.pdf.

A variation of pruning commonly practiced in water-efficient landscapes is deadheading to encourage prolonged bloom of perennial flowering plants. Deadheading prolongs bloom by removing spent flower blossoms before they mature into seedheads. Depending on the number of perennial plants in a landscape, this can be a significant maintenance task requiring a lot of time. Deadheading is not critical to optimal performance of plants in perennial beds, but may be important if season-long color is desired. Another strategy for prolonging seasonal color is to plant beds with some perennials that bloom early and others that bloom late in the season. This results in season-long color without the need for increased labor due to extensive deadheading (see "Principle 3: Appropriate Plant Selection" for bloom times of perennial flowering plants).

Many herbaceous perennials that have become weakened by age or overgrowth may need to be *divided*. A sign that plants may need dividing is when flowering is reduced, and the stems are thin and nonproductive. Dividing means digging up the plant and its roots by using a shovel, two garden forks, or a sharp blade, and prying the roots apart at natural breaks in the crown into two or more plantlets. New plants that result from division may be shared with friends or neighbors, composted, or used to fill in gaps in the landscape. Most perennials need to be divided every three to five years. An alternative to dividing is to replace aging perennials with new plants; this may be the perfect time to try something new, or to make changes to the theme of a flower bed.

Overall, water-efficient gardens are more adapted to the regional environment than conventional landscapes and should require less maintenance. However, this may not always be the case. For example, a garden with many perennials and limited turfgrass might require more maintenance time for deadheading than a conventional landscape would require for mowing, although lawns generally require greater inputs in water and fertilization. Landscapes emphasizing trees and shrubs may use fewer resources than either perennial or conventional turfgrass landscapes because of the reduced maintenance requirements. If the concept of water-efficient landscaping is expanded to embrace sustainable landscaping, then the overall landscape plan should address ways to reduce labor, fertilizer, fuels, and pesticides in addition to reducing irrigation requirements. Remember that a water-efficient landscape is not necessarily a low-maintenance landscape; rather, maintenance is a function of the landscape design and goals.

References and Additional Resources

Cox, L., and R. Koenig. 2003. "Solutions to soil problems: II. High pH (alkaline soil)." Utah State University Cooperative Extension. http://extension.usu.edu/files/publications/publication/AG_Soils_2003-02.pdf.

Beddes, T., and H. A. Kratsch. 2008. "Water-wise landscaping: Plant maintenance." Utah State University Cooperative Extension. http://extension.usu.edu/files/publications/publication/HG_Landscaping_2008-01pr.pdf.

Booth, Norman K., and James E. Hiss. 2002. *Residential landscape architecture: Design process for the Private Residence*. 3rd ed. Englewood Cliffs, NJ: Prentice Hall.

Ellefson, C. L., and D. Winger. 2004. *Xeriscape Colorado: The complete guide*. Englewood, CO: Westcliffe Publishers.

Farrell-Poe, K., J. Barnhill, R. Koenig, and B. Miller. 1997. "Using compost in Utah gardens." Utah State University Cooperative Extension. http://extension.usu.edu/files/publications/factsheet/HG_Compost_02.pdf.

Farrell-Poe, K., R. Koenig, B. Miller, and J. Barnhill. 1997. "Using compost in Utah turf applications." Utah State University Cooperative Extension. http://extension.usu.edu/files/publications/factsheet/HG_Compost_03.pdf.

Hannebaum, L. G. 2002. *Landscape design: A practical approach*. 5th ed. Englewood Cliffs, NJ: Prentice-Hall.

Heaton, K., and R. Koenig. 2003. "Solutions to soil problems: V. Low organic matter." Utah State University Cooperative Extension. http://extension.usu.edu/files/publications/publication/AG_Soils_2003-05.pdf.

Johnson, M., and R. Koenig. 2003. "Solutions to soil problems: III. Drainage." Utah State University Cooperative Extension. http://extension.usu.edu/files/publications/publication/AG_Soils_2003-03.pdf.

Keane, T. 1995. "Water-wise landscaping: Guide for water management planning." Utah State University Cooperative Extension. http://extension.usu.edu/files/publications/publication/EC_458.pdf.

Koenig, R., and T. Cerny. 2003. "Solutions to soil problems: IV. Soil structure (compaction)." Utah State University Cooperative Extension. http://extension.usu.edu/files/publications/publication/AG_Soils_2003-04.pdf.

Koenig, R., and V. Isaman. 1997. "Topsoil quality guidelines for landscaping." Utah State University Cooperative Extension. http://extension.usu.edu/forestry/Reading/Assets/PDFDocs/AG/AGSO02.pdf.

Koenig, R., K. Kopp, and C. Reid. 2002. "Water-wise landscaping: Monitoring irrigation with probes." Utah State University Cooperative Extension. http://extension.usu.edu/files/publications/publication/HG-518.pdf

Kopp, K., and P. Johnson. 2006. "Turfgrass Cultivation (Aerification)." Utah State University Cooperative Extension. http://extension.usu.edu/files/publications/publication/HG_Turf_2006-02.pdf.

Mee, W., J. Barnes, R. Kjelgren, R. Sutton, T. Cerny, and C. Johnson. *Water Wise: Native Plants for Intermountain Landscapes*. Logan, UT: Utah State University Press.

Parent, V., and R. Koenig. 2003. "Solutions to soil problems: I. High salinity (soluble salts)." Utah State University Coöperative Extension. http://extension.usu.edu/files/publications/publication/AG_Soils_2003-01.pdf.

Pittenger, D. R., D. A. Shaw, D. R. Hodel, and D. B. Holt. 2001. Responses of landscape groundcovers to minimum irrigation. *J. Environ. Hort.* 19(2):78–84.

Rumary, M. 2001. *Xeriscaping: Planning and planting low-water gardens*. New York: Sterling Publishing.

Shaw, D. A., and D. R. Pittenger. 2004. Performance of landscape ornamentals given irrigation treatments based on reference evapotranspiration. *Acta Hort.* 664:607–613.

Sowell, J. 2001. *Desert Ecology*. Salt Lake City, UT: The University of Utah Press.

Staats, D., and J. E. Klett. 1995. Water conservation potential and quality of non-turf groundcovers versus Kentucky bluegrass under increasing levels of drought stress. *J. Environ. Hort.* 13:181–185.

Weinstein, G. 1999. *Xeriscape handbook: A how-to guide to natural, resource-wise gardening*. Golden, CO: Fulcrum Publishing.

Welsh, D. F. 2001. Refining the concept of xeriscape. In *Water right: Conserving our water, preserving our environment*, ed. H. M. Cathey, 39–41. Rolling Meadows, IL: International Turf Producers Foundation. http://www.thelawninstitute.org/img/c/f185578/Water_Right_Book.PDF.

Index

Page numbers in italics refer to checklists, figures, and tables.

R

S

X